Fabrication Station

NATIONAL APPRENTICESHIP TRAINING PROGRAM FOR COOKS

AMERICAN TECHNICAL PUBLISHERS
ORLAND PARK, ILLINOIS 60467-5756

American Culinary Federation

The *National Apprenticeship Training Program for Cooks Fabrication Station* module contains procedures commonly practiced in the foodservice industry. Specific procedures vary with each task and must be performed by a qualified person. For maximum safety, always refer to specific manufacturer recommendations, insurance regulations, specific job site procedures, applicable federal, state, and local regulations, and any authority having jurisdiction. The material contained is intended to be an educational resource for the user. American Technical Publishers, Inc. assumes no responsibility or liability in connection with this material or its use by any individual or organization.

American Technical Publishers, Inc., Editorial Staff

Editor in Chief:
        Jonathan F. Gosse
Vice President—Production:
        Peter A. Zurlis
Director of Product Development:
        Cathy A. Scruggs
Art Manager:
        James M. Clarke
Multimedia Manager:
        Carl R. Hansen
Technical Editors:
        Cathy A. Scruggs
        Larry E. Pierce
Copy Editor:
        Catherine A. Mini
Cover Design:
        Jennifer M. Hines
Illustration/Layout:
        Jennifer M. Hines
        Melanie G. Doornbos
        Samuel T. Tucker
CD-ROM Development:
        Gretje Dahl
        Nicole S. Polak
        Daniel Kundrat

2 3 4 5 6 7 8 9 – 12 – 9 8 7 6 5 4 3 2

Printed in the United States of America

        ISBN 978-0-8269-4189-3

 This book is printed on recycled paper.

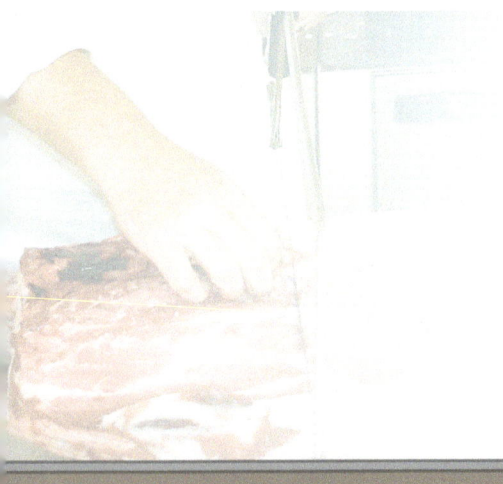

Acknowledgments

The American Culinary Federation Education Foundation (ACFEF) apprenticeship program and all its success are the result of the efforts of many individuals who work for the betterment of our apprentices and our organization as a whole. First and foremost, we thank the apprentices. Without their dedication and passion for becoming the next generation of professional chefs, we would have no goal to strive for. They work long hours in tough environments to become better culinarians. Their determination does not go unnoticed and is often what fuels the rest of us.

Just as importantly, we acknowledge the steadfast commitment of our many supervising chefs and program coordinators. They dedicate countless hours of time and effort to enriching the education of numerous apprentices, often without additional compensation or recognition. They tirelessly lead a new generation of culinarians to professional enlightenment.

No acknowledgment would be complete without mention of the many volunteers who make up the ACFEF National Apprenticeship Committee. These volunteers were invaluable in crafting and honing the current ACFEF apprenticeship program. Their time commitment was exceeded only by the value of their combined expertise. Without them, none of this would be possible.

American Technical Publishers and the American Culinary Federation sincerely appreciate the content expert who assisted with the development of this module.

Michael J. McGreal, CEC, CCE, CHE, FMP, CHA, MCFE
Department Chair, Culinary Arts & Hospitality Management
Joliet Junior College

Michael S. Baldwin, CCE
Chef Instructor
Bellingham Technical College

American Technical Publishers and the American Culinary Federation also appreciate the experts who reviewed this module.

Kevin C. Clarke, CCE, J.D.
Professor of Culinary Arts
Director of Culinary Education
Colorado Mountain College Culinary Institute

SFC Ronald Stafford
Enlisted Aide Instructor
Advance Culinary Instructor
92G Apprenticeship Program Coordinator

Elizabeth Dickson, CEPC, CCI
Culinary Arts Director
Bossier Parish Community College

American Technical Publishers and the American Culinary Federation are grateful for the technical assistance and images provided by the following companies, organizations, and individuals:

Agricultural Research Service, USDA
American Lamb Board
Canadian Beef, Beef Information Centre
Carlisle FoodService Products
Czimer's Game & Seafoods, Inc.
Daniel NYC
D'Artagnan, Photography by Doug Adams Studio
Dexter-Russell, Inc.
Edlund Co.

Florida Department of Agriculture and Consumer Services, Bureau of Seafood and Aquaculture Marketing
Fortune Fish Company
Harbor Seafood, Inc.
National Cattlemen's Beef Association
National Oceanic and Atmospheric Administration
National Pork Producers Council
National Turkey Federation

New Zealand Greenshell™ Mussels
Paderno World Cuisine
Perdue Foodservice, Perdue Farms Incorporated
Plitt Seafood
Strauss Free Raised
United States Department of Agriculture
U.S. Fish & Wildlife Service
U.S. Wellness Meats

Contents

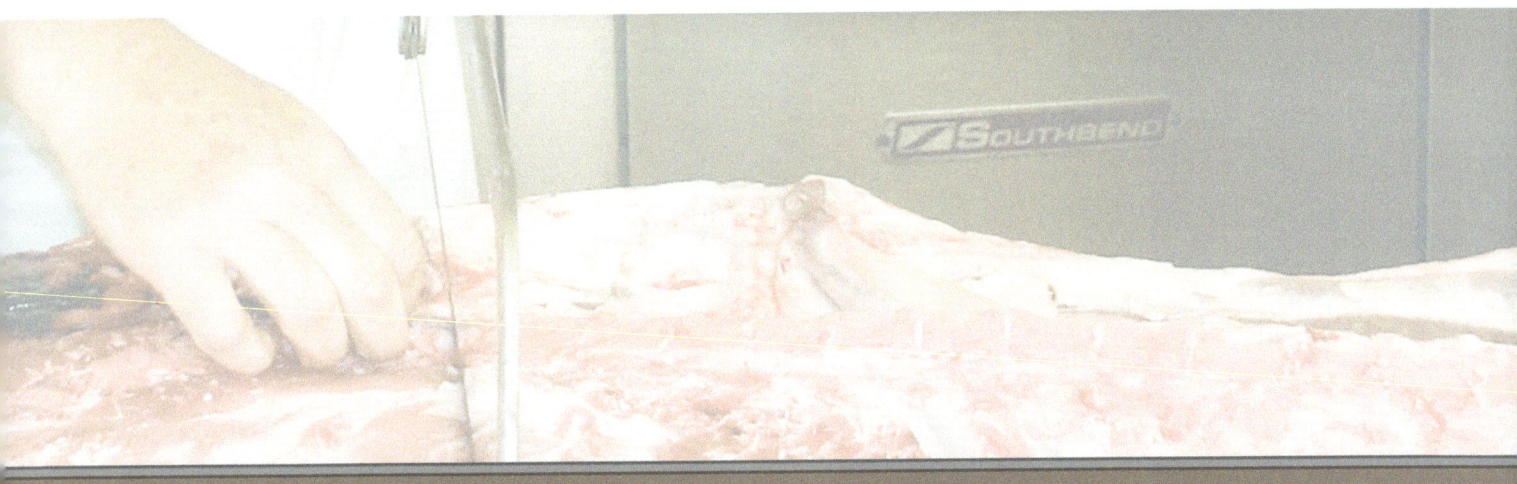

Contents

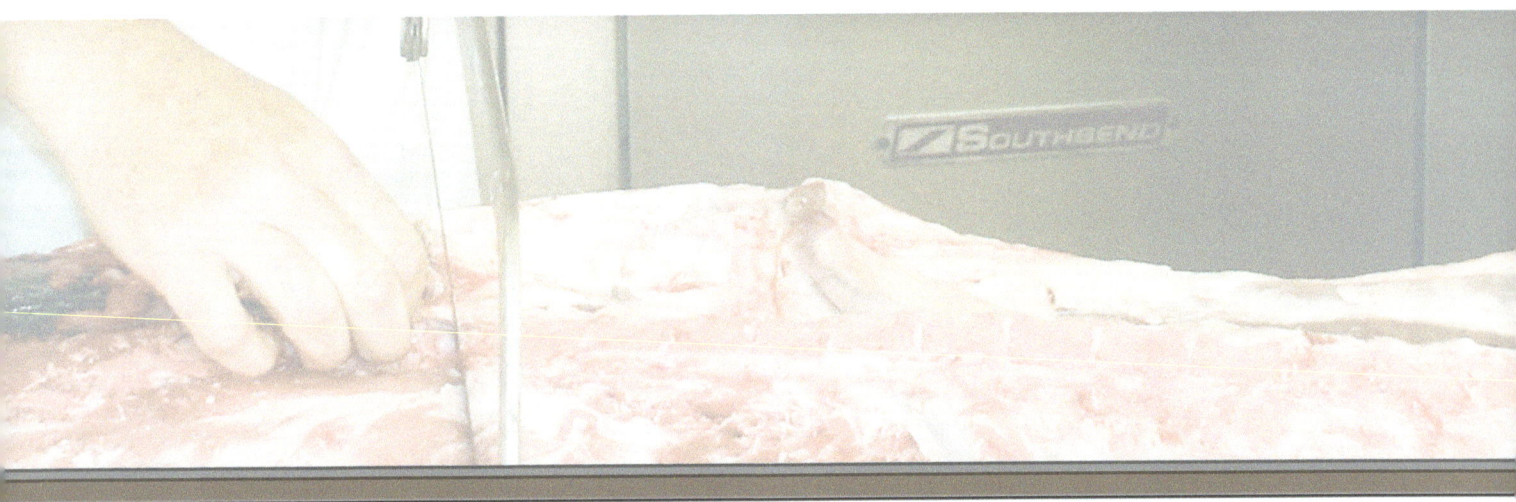

INTERACTIVE CD-ROM CONTENTS

- *Quick Quiz®*
- *Illustrated Glossary*
- *Flash Cards*
- *Media Clips*
- *Checkpoints*

- *Culinary Math Applications*
- *Certification Exam Preparation*
- *Apprenticeship Online Portal*
- *ATPeResources.com*

Introduction

Welcome to the American Culinary Federation Education Foundation (ACFEF) apprenticeship program. Whether you are registered as an ACFEF apprentice or have a thirst for learning more, the culinary techniques outlined in this book have been validated by the culinary industry and sanctioned by the U.S. Department of Labor and the ACFEF National Apprenticeship Committee.

The ACFEF apprenticeship program focuses on hands-on training that allows apprentices to learn while being mentored by leading chefs in the industry. The program is rigorous and requires apprentices to complete 445 hours of classroom instruction with a minimum of 4,000 working hours in the kitchen. From the first to last day of the apprenticeship program, ACFEF apprentices are fully immersed in the smells, tastes, and textures of their culinary creations.

Several different options for the apprenticeship program are offered to meet the specific needs of apprentices. Program options range from two to three years and include the required classroom hours. Training is organized into 10 different stations, allowing apprentices to work at their own pace. It is an "earn while you learn" approach to training. Graduates from the ACFEF apprenticeship program earn recognition as an apprentice from the U.S. Department of Labor and the American Culinary Federation (ACF) and are given the opportunity to test for industry-recognized certification.

The ACFEF apprenticeship program also addresses sustainability. The quality of the food a chef serves is directly related to the health of the ecosystem. The program encompasses a multitude of topics from recycling and composting to purchasing locally and organically.

The journey to becoming a chef can be challenging and at times may seem difficult. But there is no substitute for valuable experience. Best of luck to you as you embark on your way to becoming a chef.

The ACFEF Apprenticeship Team

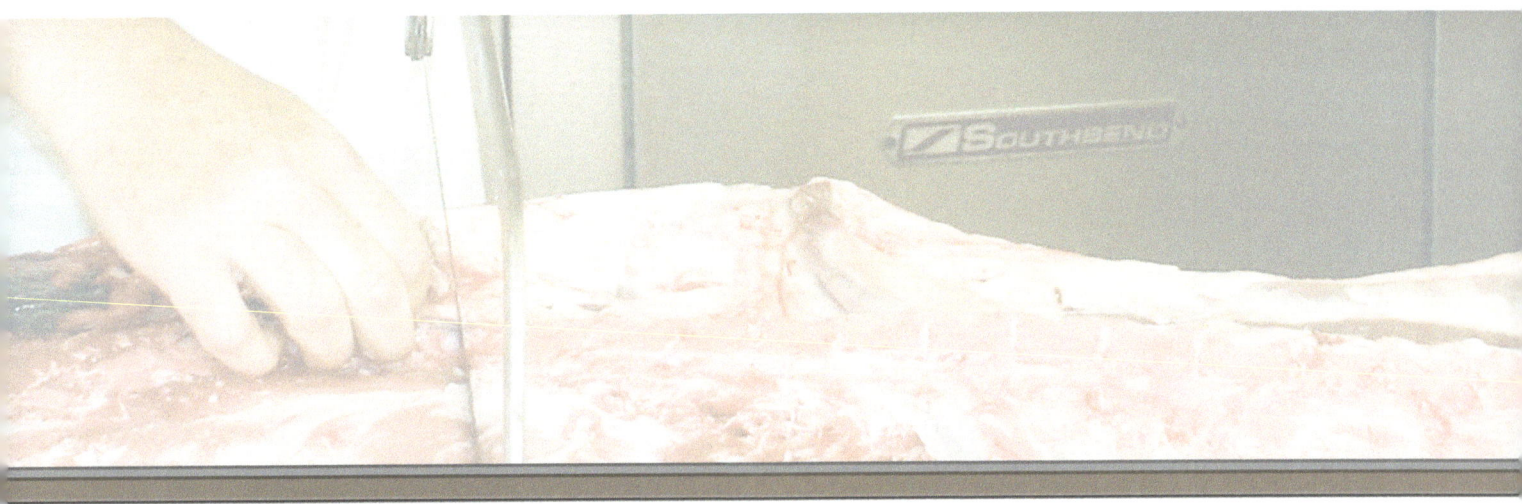

THE FABRICATION STATION

In the fourth ACFEF apprenticeship station, you will learn fabrication techniques for poultry, fish, shellfish, and various meats. As an apprentice, you will learn to identify the different varieties and market forms of these proteins. Product identification is essential to learning the step-by-step fabrication of poultry, fish, shellfish, beef, veal, pork, and lamb. Fabricating beef will include techniques for turning a fillet of beef into Chateaubriand, filets mignons, and tournedos. Proteins such as veal, pork, and lamb will be broken down from primal cuts into scallopini, chops, and racks. You will also learn how to fabricate a rabbit.

In this station, you will practice knife-sharpening techniques using a steel grinder, stone grinder, and electric grinder. You will also learn proper handling, safety, sanitation, and storage methods for fabrication equipment, including bone saws and meat grinders.

As you advance your knowledge of fabrication, you will also learn about "food miles" and other sustainable practices, such as identifying protein products that can be purchased locally and procuring a vendor for these products. Understanding the concept of sustainable proteins and identifying the benefits of purchasing locally will further increase your use of best practices in food production.

American Culinary Federation
Education Foundation

Features

MODULE FEATURES

The *National Apprenticeship Training Program for Cooks Fabrication Station* module includes several features to make learning easier.

Checkpoints appear throughout the module, and each one serves as a comprehension check of a specific portion of the module content.

Nutrition Notes highlight key nutrition information about specific foods.

Certification Exam Preparation Questions, located at the end of the module, include 20 multiple-choice items, 4 essay items, and 1 sketching activity.

Sustainability Corner addresses content related to topics such as recycling, composting, energy conservation, and product repurposing. A QR code is included with this feature.

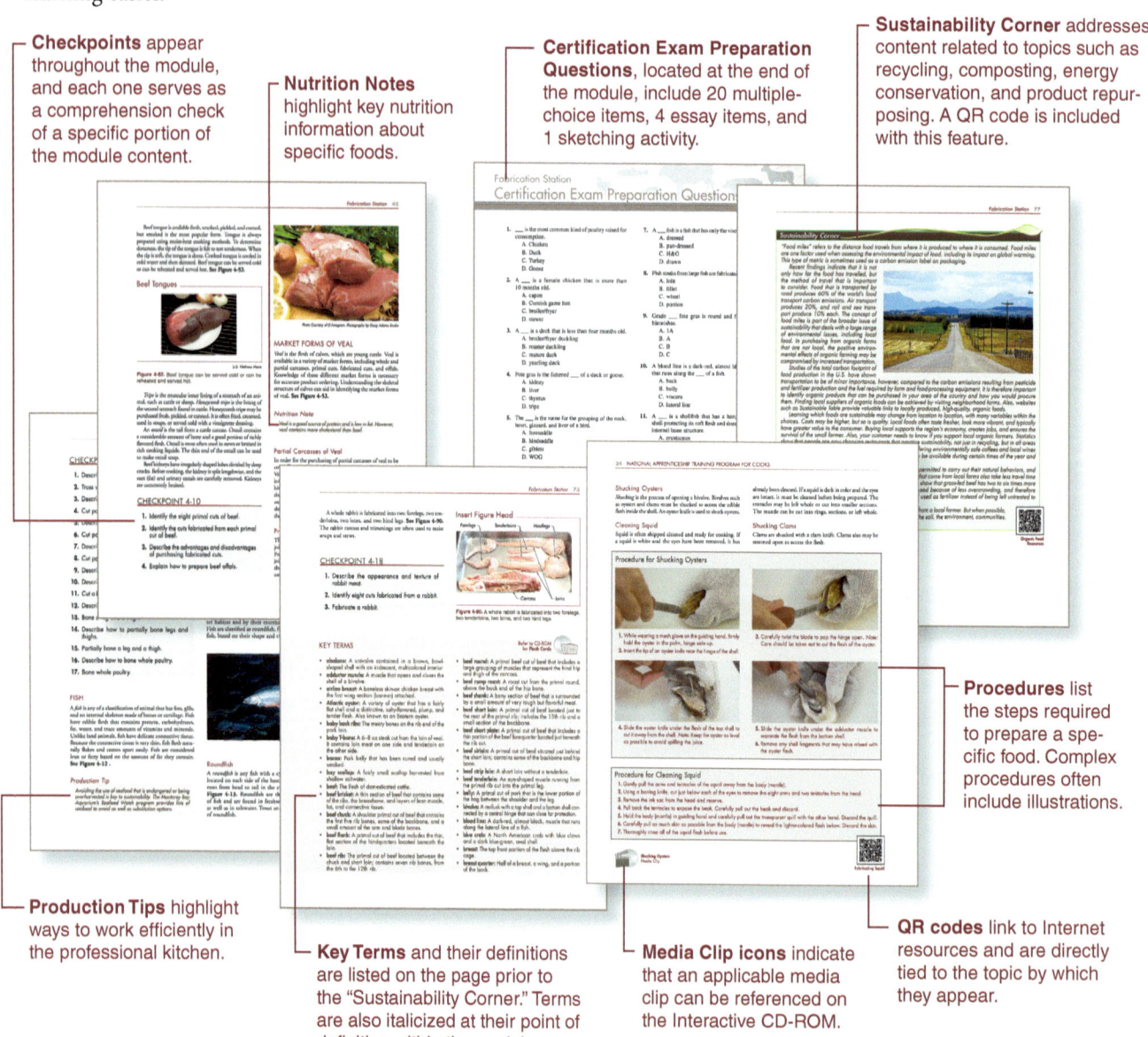

Production Tips highlight ways to work efficiently in the professional kitchen.

Key Terms and their definitions are listed on the page prior to the "Sustainability Corner." Terms are also italicized at their point of definition within the module.

Media Clip icons indicate that an applicable media clip can be referenced on the Interactive CD-ROM.

QR codes link to Internet resources and are directly tied to the topic by which they appear.

Procedures list the steps required to prepare a specific food. Complex procedures often include illustrations.

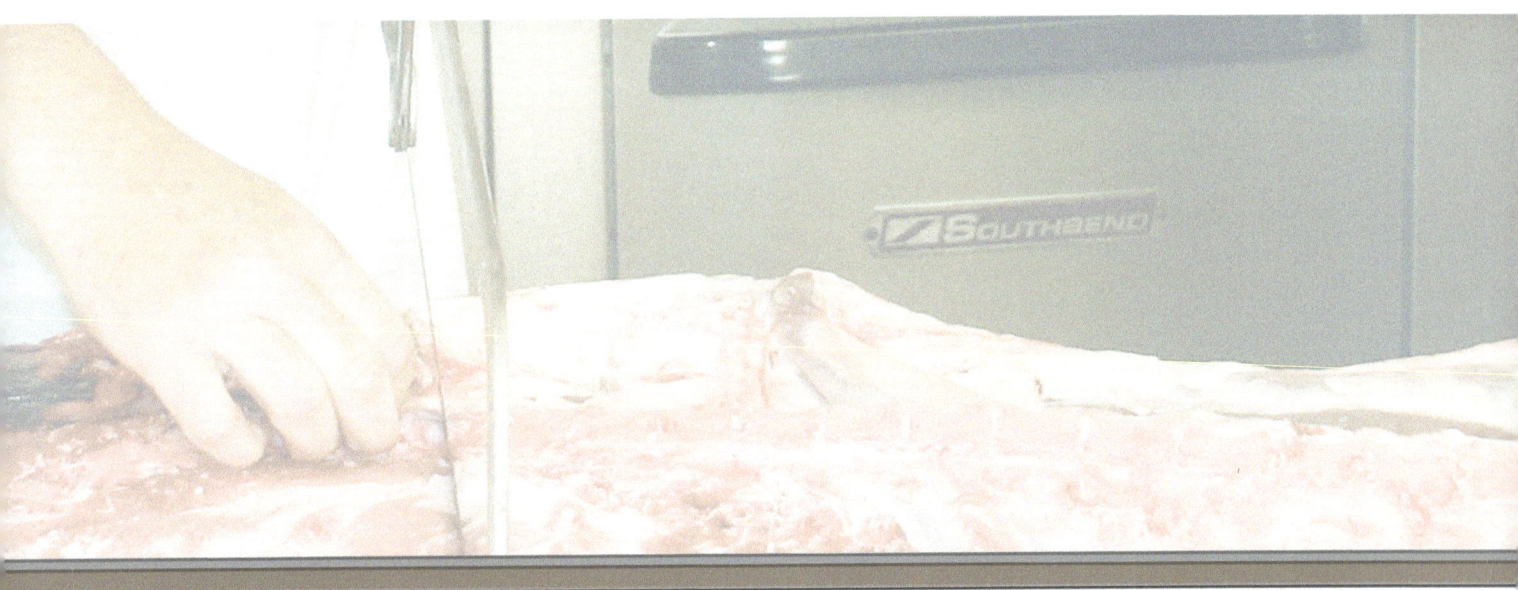

INTERACTIVE CD-ROM FEATURES

The *National Apprenticeship Training Program for Cooks Fabrication Station Interactive CD-ROM* is a self-study aid that reinforces content. This CD-ROM is Windows® compatible.

Quick Quiz® provides 25 interactive questions with embedded links to highlighted content within the module and to the Illustrated Glossary.

Illustrated Glossary provides a reference to commonly used terms. Selected terms are linked to interactive illustrations or media clips.

Flash Cards display terms and definitions, units of measure, measurement equivalents, knives, smallwares, and foodservice equipment.

Media Clips provide animations and video clips that reinforce and expand upon module content.

Checkpoints provide editable PDF files of the checkpoint review questions within the module.

Culinary Math Applications offer animations and video clips that review and reinforce math skills used in the professional kitchen.

Certification Exam Preparation provides 25 multiple-choice questions in certification exam format.

Apprenticeship Online Portal allows apprentices to track training hours, make logbook entries, review culinary resources, and build online portfolios. Supervising chefs and program coordinators can use the portal to evaluate apprentice training, grade logbooks, and communicate with apprentices.

ATPeResources.com links to online reference materials that support continued learning.

Fabrication Station

Fabrication is the skill of breaking down poultry, fish, shellfish, and meats into standardized cuts. Standardized cuts enable a chef to maximize the yield from each animal. The ability to identify each cut is useful for ordering, receiving, and storing. Also, understanding the various market forms available and the terminology used is important for efficient purchasing. Knowledge of grading also ensures that the product ordered will satisfy the chef as well as the guests.

POULTRY

Poultry is the collective term for various kinds of birds that are raised for human consumption. The kinds of poultry recognized by the United States Department of Agriculture (USDA) that are commonly served in restaurants include chickens, turkeys, ducks, and geese. Poultry is inspected by the USDA for wholesomeness. However, the grading of poultry for quality is voluntary and paid for by poultry producers and/or processors. Each kind of poultry is subdivided into classes based on age and/or gender.

Chickens

Chicken is the most common kind of poultry raised for consumption. Chickens are classified by age and sometimes gender. Common chicken classifications include Cornish game hens, broilers/fryers, roasters, capons, and stewers (stewing hens). **See Figure 4-1.**

Cornish Game Hens. A *Cornish game hen,* also known as a Rock Cornish game hen, is either a female or male chicken that is less than five weeks old. Cornish game hens typically weigh 1½ lb or less. The term "hen" is commonly reserved for female birds, but the Cornish game hen is the exception. The flesh of a Cornish game hen is very tender and mildly flavored. Cornish game hens are most often stuffed and roasted whole.

Safe Poultry Handling

Common Chicken Classifications

Cornish Game Hen

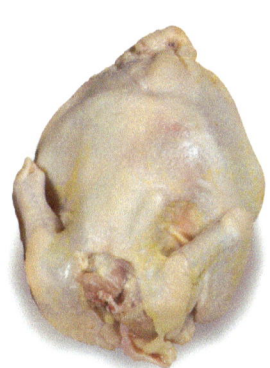

Broiler/Fryer

Stewer (Stewing Hen)

Figure 4-1. Chicken classifications include Cornish game hens, broilers/fryers, and stewers. Roasters and capons are also available.

Broilers/Fryers. A *broiler/fryer* is a young male or female chicken less than five months old. Broiler/fryers typically weigh 1½–3½ lb. Broilers/fryers have very tender flesh and smooth skin. They contain a slightly higher percentage of fat than Cornish hens. Any cooking method can be used to prepare broilers/fryers.

Roasters. A *roaster* is a young male or female chicken that is from 2–3 months old and has a ready-to-cook carcass weight of 5 lb or more. Roasters have tender flesh and smooth skin. Roasters can be prepared using any cooking method.

Capons. A *capon* is a surgically castrated male chicken that is less than four months old. Capons weigh approximately 4–7 lb. Capons are castrated to produce large, well-formed breasts with flesh that is more tender than that of broilers/fryers. Capons are most often roasted.

Stewers. A *stewer,* also known as a stewing hen, is a female chicken that is more than 10 months old. Stewers weigh approximately 3–8 lb. The flesh and skin of a stewer is tough, but flavorful due to the age of the bird. A stewer requires slow moist-heat cooking methods, such as stewing or braising.

Turkeys

Turkeys are also classified by age. Classifications of turkeys include fryer/roaster turkeys, young turkeys, yearling turkeys, and mature turkeys.

Fryer/Roaster Turkeys. A *fryer/roaster turkey* is a male or female turkey that is less than three months old. Fryer/roaster turkeys weigh approximately 4–9 lb. Fryer/roaster turkeys are very tender and have soft, flexible skin. Male fryer/roaster turkeys are commonly called "toms." Female turkeys are commonly called "hens." Common preparation methods include roasting, sautéing, and pan-frying.

Young Turkeys. A *young turkey* is a male or female turkey that is less than eight months old. Young turkeys weigh approximately 8–22 lb. Young turkeys have tender flesh, smooth skin, and a firm breastbone. Young turkeys are commonly roasted or stewed.

Yearling Turkeys. A *yearling turkey* is a mature turkey that is less than 15 months old. Yearling turkeys weigh approximately 10–30 lb. The flesh of a yearling turkey is still tender. **See Figure 4-2.** Yearlings are commonly roasted or stewed.

Mature Turkeys. A *mature turkey* is a turkey that is more than 15 months old. Like yearlings, mature turkeys range in weight from 10–30 lb. The flesh and skin of a mature turkey is tough but flavorful due to the age of the bird. Mature turkeys are often roasted or stewed.

Yearling Turkeys

Photo Courtesy of D'Artagnan
Photography by Doug Adams Studio

Figure 4-2. The flesh of a yearling turkey is still tender.

Ducks

Ducks are classified by age. Classifications of duck commonly served in restaurants include broiler/fryer ducklings, roaster ducklings, and mature ducks.

Broiler/Fryer Ducklings. A *broiler/fryer duckling* is a duck that is less than two months old. Broiler/fryer ducklings weigh approximately 3–6½ lb. Broiler/fryer ducklings have very tender flesh and a soft windpipe. They are commonly roasted or stewed.

Roaster Ducklings. A *roaster duckling* is a duck that is less than four months old. Roaster ducklings weigh approximately 4–7½ lb. The flesh of a roaster duckling is tender, and the windpipe is just starting to harden. Roaster ducklings are most often roasted.

Mature Ducks. A *mature duck* is a duck that is more than six months old. The flesh of a mature duck is fairly tough and the windpipe is hardened. The flesh of mature ducks is typically used in processed duck products. Mature ducks are typically braised.

White Pekin ducks, Muscovy, Moulard, and Mallard ducks are sold for consumption. The Mallard duck is the breed from which all domestic ducks, except the Muscovy, are descended. Farm-raised Mallards are available on a limited basis. Each breed of duck is known for one or more products or uses. **See Figure 4-3.**

• The white Pekin duck, also known as Long Island duckling, is the most popular breed of duck consumed in the United States. The white Pekin is typically used for roast duck. This breed of duck should not be confused with the dish named Peking duck, which consists of duck served with pancakes, scallions, and hoisin sauce.

• The Muscovy duck, also known as the Barbary or Barbarie duck, has thinner skin and much less fat than other breeds of duck. Male Muscovy ducks are meatier than the females. Muscovy flesh is redder and has a more pronounced flavor than other ducks. Each side of a Muscovy breast looks like a "duck steak" in that it weighs approximately 8 oz. The liver of the Muscovy is also sold, but it is not as preferred as Moulard liver.

- The Moulard duck is a sterile hybrid of a male Pekin duck and a female Muscovy duck. The Moulard is larger and considerably fatter than the white Pekin duck and the flesh has a strong gamey flavor. Moulards are bred for their exceptional livers and breast meat. The breast of a Moulard is called a magret and must be served rare in order to be tender.
- The Mallard duck is smaller in size and its flesh is tougher than the white Pekin duck. The flesh of a Mallard has very little fat, yet it can taste greasy and gamey.

Ducks have less flesh in proportion to bone and fat than most other kinds of poultry. For example, a duck yields half as much flesh as a chicken of the same size.

The majority of fat in a duck is located in and just beneath the skin. With the fatty skin removed, duck is lean in comparison to other kinds of poultry. When cooking duck, the fatty skin must be rendered slowly or it becomes nearly inedible.

Duck breasts are often seared in a sauté pan, finished in an oven, and served medium-rare. Duck legs are most often braised or roasted until well-done and tender. Whole ducks are usually roasted on a rack to allow the fat to render.

Duck legs and thighs may be prepared as confit. *Confit* is a French term for meat that has been cooked and preserved in its own fat. To make a duck confit, pieces of duck leg and thigh are salted and then simmered in rendered duck fat until tender. The confit can be served whole, or the flesh can be pulled from the bone and used as a flavorful ingredient in salads or other dishes. A *cassoulet* is a dish that consists of white beans stewed with duck fat, fresh sausage, and duck confit.

Foie gras is the fattened liver of a duck or goose. **See Figure 4-4.** Foie gras is a delicacy that is considered a staple of classic fine dining. A duck or goose that is used to make foie gras is fed a rich diet until the liver becomes almost solid fat. The fattened liver is removed and sold as the most expensive part of the duck or goose.

Foie Gras

*Photo Courtesy of D'Artagnan,
Photography by Doug Adams Studio*

Figure 4-4. Foie gras is the fattened liver of a duck or goose.

Foie gras is graded based on the size, appearance, and texture of the liver. To receive a grade of A, the liver must weigh at least 1 lb 8 oz and be round and firm with no blemishes. Grade A foie gras is used for terrines and pâtés. Grade B foie gras weighs between 1 lb and 1 lb 7 oz, has a good texture, but is less round in shape than Grade A foie gras. Grade B is a good choice for poaching or sautéing. Foie gras that weighs less than 1 lb, is slightly flattened, and has some visual imperfections and soft spots is grade C. Grade C foie gras is used primarily for mousses and pâtés.

Whole Ducks and Duck Breasts

Pekin Duck

Muscovy Duck

Duck Breast

*Photos Courtesy of D'Artagnan,
Photography by Doug Adams Studio*

Figure 4-3. Common market forms of duck include whole ducks and duck breasts.

Because foie gras is almost all fat, it is typically seared in a very hot sauté pan and served immediately. Foie gras may also be poached, cooled, and puréed to make a pâté de foie gras.

Geese

Geese have dark flesh and a large amount of fat in both the skin and the flesh. A goose is classified as either young or mature. Only young geese are used in foodservice operations.

A *young goose* is a goose that is usually less than six months of age and weighs approximately 4–10 lb. **See Figure 4-5.** Young geese have tender flesh and a windpipe that is easily dented. Their flesh has a rich taste due to the high fat content. The large flaps of thick, fatty skin around the neck and tail should be removed with a knife or kitchen shears prior to cooking. A young goose is commonly roasted at very-high temperatures to aid in rendering some of the fat from the skin and the flesh.

Goose

Czimer's Game & Seafoods, Inc.

Figure 4-5. A young goose weighs approximately 4–10 lb.

CHECKPOINT 4-1

1. Identify two traits that are used by the USDA to classify poultry.

2. Describe five classifications of chicken.

3. Describe four classifications of turkey.

4. Describe three classifications of duck.

5. Differentiate between white Pekin, Muscovy, Moulard, and Mallard ducks.

6. Contrast the three grades of foie gras.

MARKET FORMS OF POULTRY

Poultry is sold after the head, neck, feet, and feathers have been removed from the bird. Poultry can be sold fresh or frozen, whole or cut-up, boneless or bone in, ground, or processed into a prepared form such as chicken tenders. Foodservice operations purchase all market forms of poultry.

Some poultry has both light and dark flesh. Other poultry has only dark flesh. Leg and thigh flesh is always dark, but breasts and wings can be either light or dark. For example, chickens spend most of their time on their feet and therefore have dark flesh in their legs and thighs. Ducks have dark flesh throughout their entire bodies, because all of their muscle groups are used for slow, sustained movement as they fly. Because the breast and wing muscles get more exercise, more blood flows through them. Blood contains the protein myoglobin, which causes the darkening of muscle tissue. The more a muscle is used, the darker and more flavorful the muscle becomes.

The fat in poultry is found just beneath the skin, around the tail, and on the abdomen. **See Figure 4-6.** Because poultry does not have much intramuscular fat, it can become very dry if it is overcooked even slightly. As a bird ages, its breastbone and keel becomes less flexible and the flesh and skin toughen, intensifying the overall flavor. Therefore, younger birds are often desired for their tenderness and mild flavor.

Poultry Fat

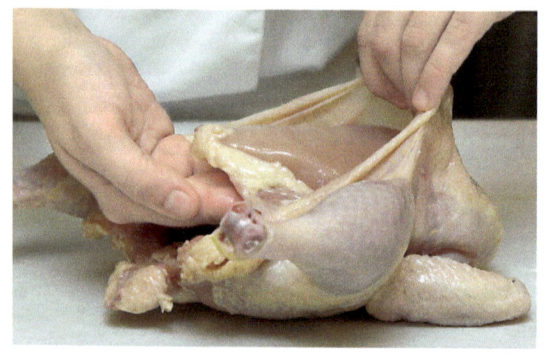

Figure 4-6. The fat in poultry is found just beneath the skin, around the tail, and on the abdomen.

Whole Poultry and Fabricated Cuts

Poultry is available in a variety of market forms, including whole birds and fabricated cuts. Knowledge of these market forms is necessary for accurate ordering. Purchasing whole poultry allows a chef to be more creative with the menu. It is also less expensive to purchase whole poultry and debone it in-house.

Fabricated cuts of poultry include breasts, tenders, tenderloins, wings, legs, leg quarters, breast quarters, halves, and ground poultry. **See Figure 4-7.** Cutlets and sausages are also available.

Whole and Fabricated Poultry Cuts

Whole Turkey

Ground Turkey

Turkey Wings

Turkey Thighs

Turkey Breast

Turkey Legs

National Turkey Federation

Figure 4-7. Poultry is sold whole and in common fabricated cuts, including ground, wings, thighs, breasts, and legs.

- A *breast* is the top front portion of the flesh above the rib cage. The breast consists of white flesh in birds that only fly in quick, short bursts and dark flesh in birds that have the ability to fly long distances.
- An *airline breast* is a boneless skin-on chicken breast with the first wing section (bone-in) attached.
- A *chicken cutlet* is a boneless, skinless section of a chicken breast that has been tenderized.
- A *tender* is a small strip of a breast.
- A *poultry tenderloin* is the inner pectoral muscle that runs alongside the breastbone of a bird.

- A *wing* consists of a tip, paddle, and drummette. A *wing tip* is the outermost section of a wing. It is often used to make stock as it contains little flesh. A *wing paddle,* also known as a wing flat, is the second section of a wing located between the two wing joints. A *drummette* is the innermost section of a wing located between the first wing joint and the shoulder.
- A *poultry leg* consists of a drumstick and thigh. A *drumstick* is the lower portion of the leg located below the hip and above the knee joint. The *thigh* is the upper section of the leg located below the hip and above the knee joint.
- A *leg quarter* is a thigh, a drumstick, and a portion of the back.
- A *breast quarter* is half of a breast, a wing, and a portion of the back.
- A *poultry half* is a full half-length of a bird split down the breast and spine.
- *Ground poultry* is ground fabricated cuts of poultry.

To help ensure that foodservice operations and suppliers communicate efficiently, the USDA publishes the Institutional Meat Purchase specifications (IMPS) for commonly purchased meats, poultry, seafood, and game products. All cuts are numbered by category.

Giblets

When purchasing whole poultry, it is common to find a small bag containing the giblets inside the cavity of the bird. *Giblets* is the name for the grouping of the neck, heart, gizzard, and liver of a bird. **See Figure 4-8.** When ordering whole poultry it is common to specify whether the giblets are desired. When purchasing poultry without giblets, the acronym WOG is used by suppliers.

Giblets

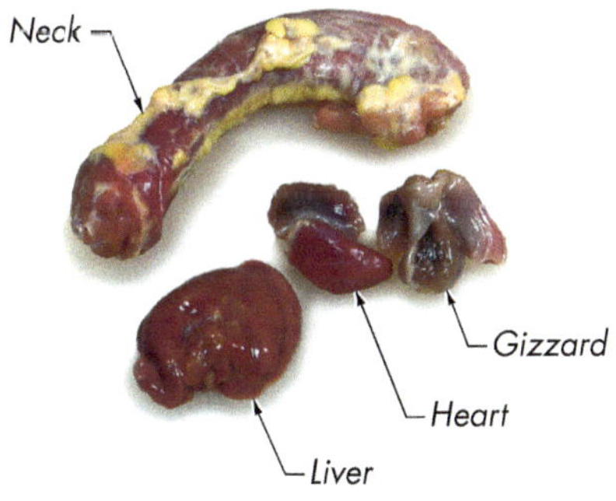

Neck

Gizzard

Heart

Liver

Figure 4-8. Giblets include the neck, heart, gizzard, and liver of a bird.

The neck, heart, and gizzard are often used to make giblet gravy. The neck contains an abundant amount of gelatin, which makes a rich and flavorful stock. Turkey hearts are served with a béchamel sauce in a classic preparation called creamed hearts. Gizzards are often trimmed of connective tissue, breaded or battered, and then fried. Livers are often breaded then fried or sautéed and served with caramelized onions. Livers are also used to make liver pâtés.

Receiving and Storing Poultry

Poultry must be stored at an internal temperature of 41°F or below. It should always be stored beneath other foods to prevent the juices from contaminating the other foods. Fresh poultry should be used within two to three days or should be frozen immediately to prevent loss of quality and potential spoilage.

Poultry spoils rapidly. Although it develops an odor as it spoils, it may be unsafe for consumption prior to developing offensive odors. Many foodservice operations specify that fresh poultry be packed in vacuumed sealed bags and sometimes gas flushed. Gas flushed bagging increases the shelf life of the poultry by replacing oxygen with nitrogen, which inhibits bacteria growth.

Frozen poultry should be frozen in its original packaging at 0°F or below. Frozen poultry should be moved to a refrigeration unit a day prior to use in order to thaw safely. Whole turkeys may need an additional one to two days to thaw because of their size. Poultry should never be refrozen once it has thawed.

Frozen convenience poultry products do not need to be thawed. Instead, frozen convenience poultry products such as chicken tenders should remain in the freezer until they are ready to be cooked. **See Figure 4-9.**

Frozen Chicken Tenders

Photo Courtesy of Perdue Foodservice, Perdue Farms Incorporated

Figure 4-9. Frozen convenience poultry products should remain in the freezer until they are ready to be cooked.

CHECKPOINT 4-2

1. Explain why some poultry has both light and dark flesh and some poultry has only dark flesh.

2. Explain the advantage of purchasing whole poultry.

3. Identify common fabricated cuts of poultry.

4. Name the items in the bag located inside whole poultry.

5. Identify special precautions to follow when receiving and storing poultry.

6. Explain how frozen poultry should be thawed.

FABRICATING POULTRY

There are many ways that poultry can be cut into portions. Typical fabrication methods for poultry include cutting them into halves, quarters, and eighths. Poultry may also be fabricated to produce boneless breasts and airline breasts. Fabricating techniques can be applied to almost any type of poultry because all types have similar bodies and bone structures. Chicken is the most economical bird to fabricate. Understanding the skeletal structure of a chicken can aid the fabrication process. **See Figure 4-10.**

Trussing Whole Poultry

When roasting a whole bird, the bird is usually trussed using butchers twine. *Trussing* is the process of tying the legs and wings of a bird tightly to the body to keep a compact shape. Trussing helps the bird cook evenly and retain moisture. Trussing also gives the bird a pleasing finished appearance. Prior to trussing, any excess fat should be trimmed from around the neck area and tail portion of the bird. The skin is then pulled tightly and evenly across the breast to cover any exposed flesh and prevent the breast from drying out during cooking. If desired, the wing tips can be removed, as they have a tendency to burn during roasting. If the wings are left intact, the first joint is tucked behind the second joint for a neat appearance.

Skeletal Structure of Chickens

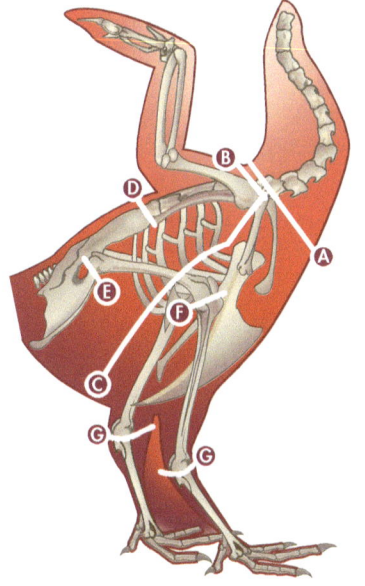

	Cut Points
A	Neck from breast/back
B	Wing from breast
C	Breast from back
D	Front half of back from back half of back
E	Leg from back
F	Thigh from drumstick
G	Drumstick from foot

Figure 4-10. Understanding the skeletal structure of a chicken can aid the fabrication process.

Procedure for Trussing Whole Poultry

1. Cut a length of butcher's twine approximately three times the length of the bird to be trussed.
2. With the breast up, place the center of the twine beneath the bird, about 1 inch under the tail.

3. Bring the twine up around the legs and cross the ends, creating an "X" between the legs.
4. Pass the ends of the twine under the legs and pull tight.

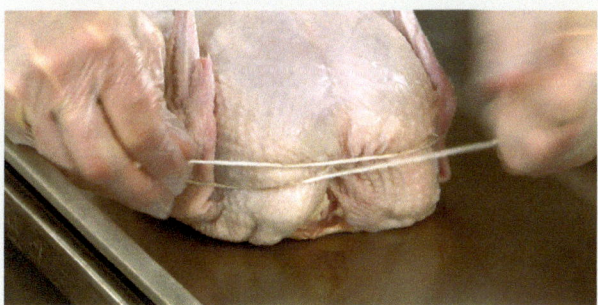

5. Turn the bird around, and pull the twine across the wings and tie a square knot at the neck to secure the truss.

Finished truss.

Trussing Poultry
Media Clip

Cutting Poultry into Halves

Poultry is commonly cut into halves. The bird is split from top to bottom between the breasts and along the backbone to the tail. This results in two equal portions.

Production Tip

Never use the same cutting boards or tools that were used with raw poultry for other products until they have been washed and sanitized.

Cutting Poultry into Quarters and Eighths

Poultry is often cut into quarters for grilling, broiling, or roasting. First, the bird is divided into leg and thigh sections and wing and breast sections. There are two in each section, yielding four quarters.

Poultry is commonly cut into eighths for pan-frying, deep-frying, grilling, broiling, and roasting. To cut a bird into eighths, quarters are cut into two breasts, two wings, two thighs, and two legs.

Procedure for Cutting Poultry into Halves

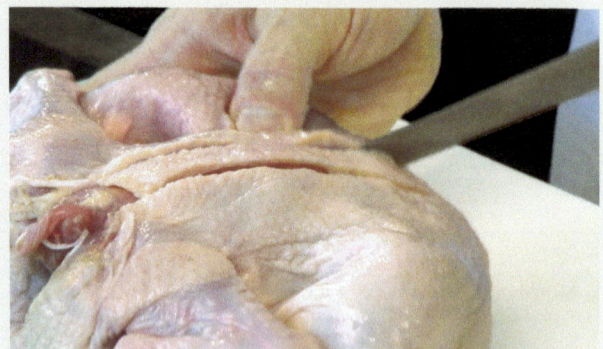

1. Square the bird by firmly squeezing the legs and wings toward the body.
2. With the breast side down, use a stiff (curved) boning knife to split the bird along both sides of the backbone from the neck to the tail.

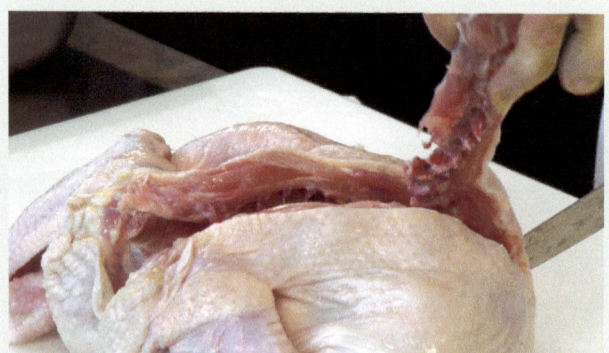

3. Remove the backbone from the carcass.

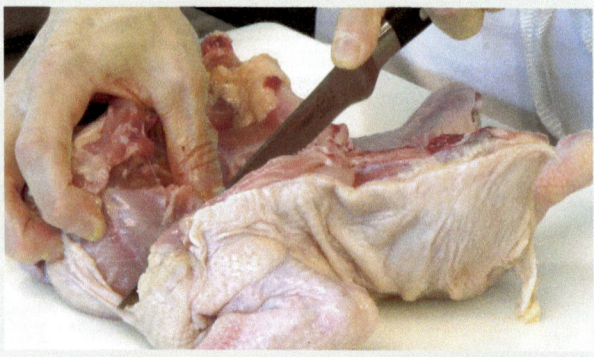

4. Open both sides of the bird to reveal the keel bone (breastbone). Cut through the keel bone and wishbone lengthwise from neck to tail. If necessary, hit the spine of the blade with the heel of the hand.
5. Cut through the flesh and skin behind the breastbone to separate the bird into halves.

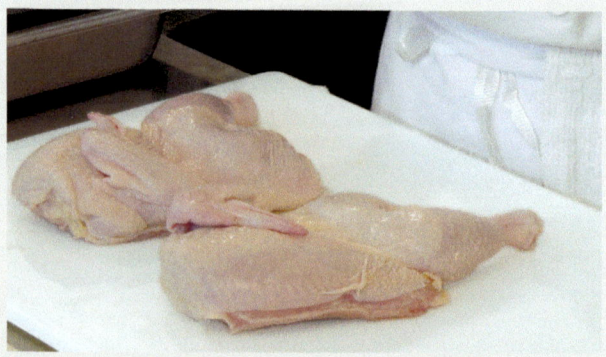

Poultry cut into halves.

Procedure for Cutting Poultry into Quarters and then Eighths

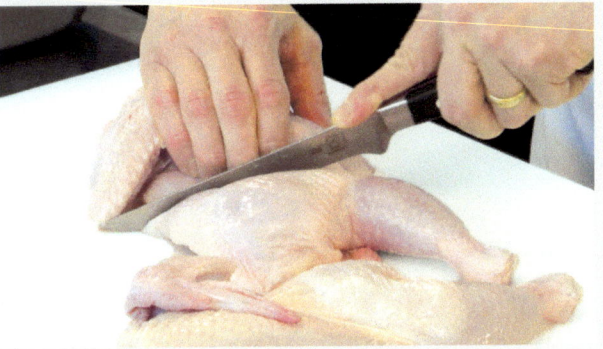

1. Cut the bird into two halves.
2. Cut through the flap of skin between the breast and thigh and pull the thigh away from the breast to expose the joint. Then, cut the joint to separate the thigh from the breast.

Poultry cut into quarters.

3. Cut the joint between the breast and the thigh to separate the breast and thigh from one half.

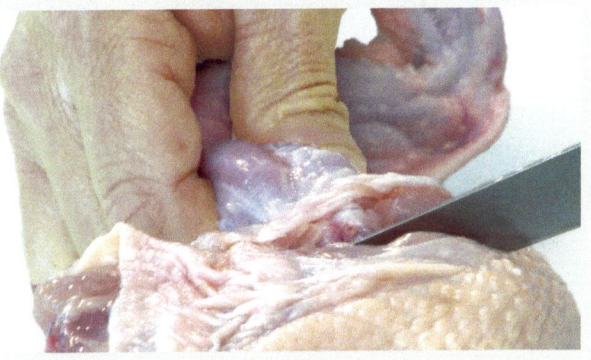

4. Hold the wing away from the breast, and cut the wing from the breast at the wing joint.

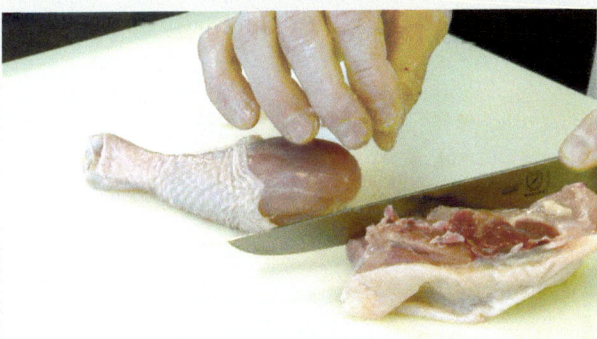

5. Hold the leg so that the inside thighbone is visible and locate the thin line of fat that separates the leg and thigh muscles. Cut along this line to separate the leg and thigh joint.

Poultry cut into eighths.

Cutting Poultry into Eighths
Media Clip

Cutting Boneless Breasts

Boneless breasts are the most popular poultry cut because of their versatility. They can be grilled, broiled, roasted, sautéed, pan-fried, poached, or stuffed. Larger breasts, such as those from a turkey, can be roasted whole or sliced into medallions and sautéed. A boneless breast is often flattened before being sautéed or stuffed. The breast is pounded flat with a meat mallet until the desired thickness is achieved. Boneless breasts are often fabricated in-house.

Procedure for Cutting Boneless Breasts

1. With the breast side down, use a stiff (curved) boning knife to split the bird along both sides of the backbone from the neck to the tail. Remove the backbone from the carcass.
2. Pull the leg and thigh away from the breast. Locate the thigh joint and cut the flesh down to the joint.
3. Twist the leg to break the thigh joint. Cut through the joint and separate the leg from the carcass.
4. Open the bird to reveal the keel bone. Cut through the keel bone and wishbone lengthwise from the neck to the tail. If necessary, hit the spine of the blade with the heel of the hand.
5. Cut through the flesh and skin behind the keel bone to separate the bird into halves.
6. Lay the bird skin-side down and cut the breast away from the rib bones. *Note:* Use care when cutting around the rib cage to keep the tenderloin attached to the breast.

Cutting Airline Breasts

Airline breasts, also known as suprêmes, are similar to boneless breasts but contain one drummette bone. **See Figure 4-11.** An airline breast makes an elegant presentation due to the exposed drummette bone.

Airline Breasts

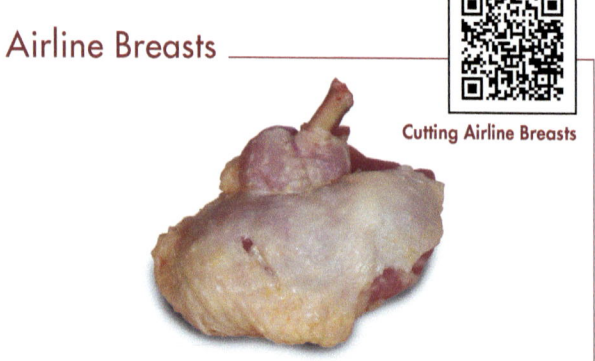

Cutting Airline Breasts

Figure 4-11. Airline breasts are similar to boneless breasts but contain one drummette bone.

Procedure for Cutting Airline Breasts

1. With the breast side down, use a stiff (curved) boning knife to split the bird along both sides of the backbone from the neck to the tail. Remove the backbone from the carcass.
2. Open the bird to reveal the keel bone. Cut through the keel bone and wishbone lengthwise from the neck to the tail. If necessary, hit the spine of the blade with the heel of the hand.
3. Cut through the flesh and skin behind the keel bone to separate the bird into halves.
4. Cut through the skin between the breast and the thigh. Pull the thigh away from the breast to expose the joint.
5. Cut the joint to separate the breast from the thigh.
6. Cut along one side of the breastbone, following the curve of the ribs, to separate the flesh from the bone.
7. Separate the wing from the rib cage by cutting the joint. Keep the wing attached to the breast.
8. Cut the breast meat free from the carcass.
9. Make a cut on the back of the joint between the drummette and the paddle bones.
10. Break the joint and pull back the flesh and skin to expose the drummette bone. Trim the end of the drummette bone to remove the cartilage. *Note:* If desired, the skin can be removed.

Boning Legs and Thighs

Legs and thighs have more flavor than breasts because they have dark flesh and additional fat. Boneless legs and thighs are often stuffed and roasted. When stuffing and roasting legs and thighs, the bones are removed but the flesh is left intact.

Boneless legs and thighs may be cut into smaller pieces for use in stir-fries and soups. They can also be flattened prior to being sautéed or stuffed. To flatten a boneless leg or thigh, the boneless leg or thigh is wrapped in plastic and then pounded flat with a meat mallet until the desired thickness or diameter is achieved. Partially boned legs and thighs are often desired to create a more elegant presentation.

Procedure for Boning Legs and Thighs

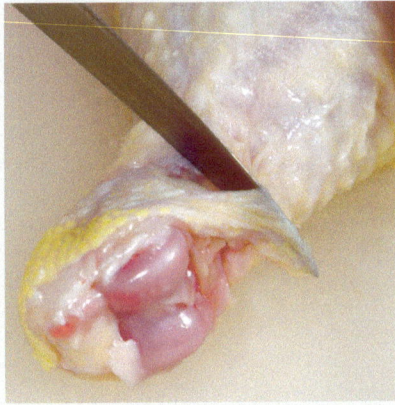

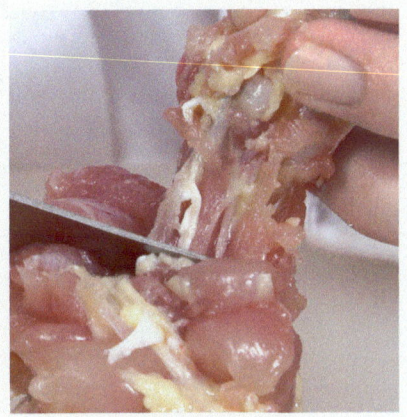

1. Place a leg quarter on a cutting board with the inside of the thigh facing upward.
2. Use a stiff (curved) boning knife to cut down the length of the thighbone and leg bone on each side and around the cartilage of the joint to free the bones from the flesh.

3. Pull the flesh away from the leg bone and cut the flesh off where it connects to the end joint.

4. With smooth, even strokes cut around the joint at the end of the leg until the L-shaped leg, thighbones, and cartilage are free from the flesh. Reserve the bones for stock.
5. Repeat with the other leg quarter. If desired, remove the skin.

Procedure for Partially Boning Legs and Thighs

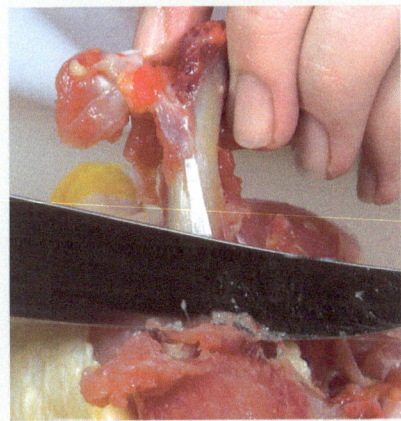

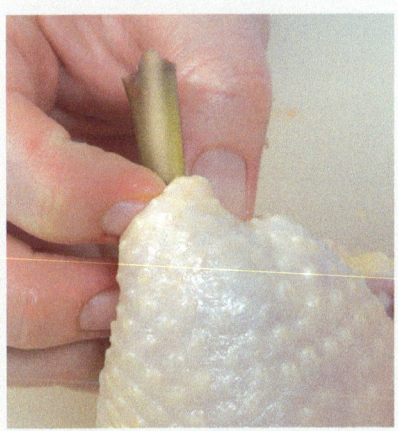

1. Use a stiff (curved) boning knife to cut down the length of the thighbone of a leg quarter.
2. Scrape the thigh flesh off the thighbone down to the joint.

3. Cut through the joint that connects the leg to the thighbone to remove the thighbone.
4. Use a chef's knife to chop off the joint.

5. Push the flesh away from the joint for a finished presentation.

Fully boned and partially boned legs and thighs may be stuffed. They are typically trussed or wrapped in caul fat before cooking to help to maintain the shape of the flesh.

Caul fat is a meshlike fatty membrane that surrounds sheep or pig intestines. Caul fat adds additional moisture and maintains the shape of the poultry being roasted.

Caul fat melts almost completely away during the cooking process, but not before it helps set the shape of the item being cooked. Stuffed legs and thighs can be seared in a sauté pan and finished in the oven, or they can be roasted or braised.

Boning Whole Poultry

Boning a whole bird is one of the more difficult fabrication techniques performed in the professional kitchen.

In this technique the bones of the bird are removed while keeping the flesh intact. Small birds, such as quail, squab, and poussins, can be boned or partially boned for a variety of preparations and presentations. Larger birds, such as chicken, are boned for charcuterie preparations. A deboned chicken is often called a butterflied chicken because it can be laid flat on a grill and will cook evenly.

Procedure for Boning Whole Poultry

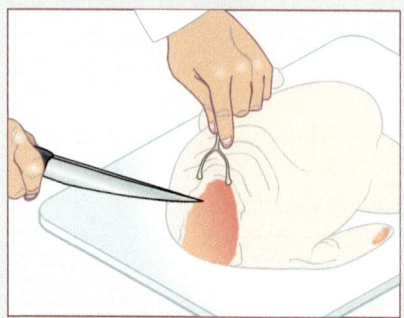

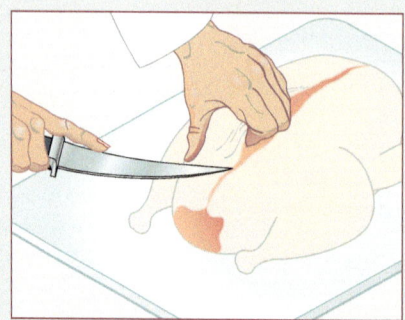

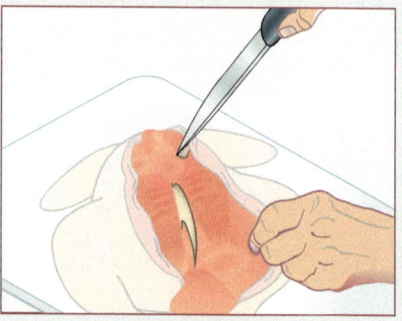

1. Place the bird breast-side up and stretch each wing flat against the cutting board by pulling on the tip. Use a stiff (curved) boning knife to cut off the wing tip and the next joint, leaving the drummette still attached.

2. Pull the skin of the neck area out of the way, and slide the knife along the underside of the wishbone. Cut around and under the wishbone until it is free and can be pulled out by hand.

3. Turn the bird breast-side down. Use short strokes to cut along the backbone from the neck to the tail, keeping the knife close to the bones. Carefully pull the flesh away from the carcass as the flesh is cut.

4. Cut through the ball-and-socket joints connecting each wing and thighbone to the carcass to separate them from the carcass while leaving them attached to the skin. *Note:* When this step is complete on both sides, the flesh will only be attached along the ridge of the breastbone.

5. Gently pull to separate the breastbone and carcass from the flesh.

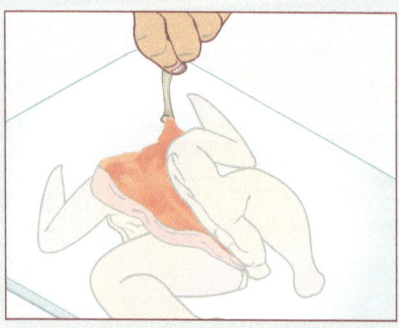

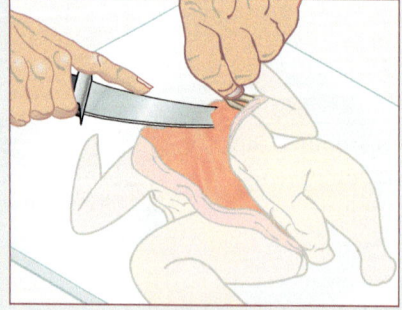

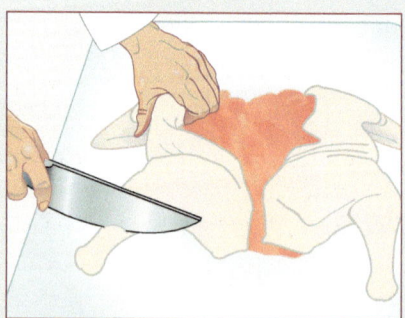

6. Cut the flesh from the curved bone near each wing to remove the bone.

7. Cut down the length of the thighbone and scrape the flesh down to the joint.

8. Cut between the leg and thighbones at the joint to remove the thighbone from each side.

9. While holding the exposed portion of the wing bone, cut through the tendons and scrape the meat from the bones.

10. Pull out the wing bone and repeat on the other side.

11. Use a chef's knife to cut off the end of the leg bone.

CHECKPOINT 4-3

1. Describe how to truss whole poultry.

2. Truss whole poultry.

3. Describe how to cut poultry into halves.

4. Cut poultry into halves.

5. Describe how to cut poultry into quarters.

6. Cut poultry into quarters.

7. Describe how to cut poultry into eighths.

8. Cut poultry into eighths.

9. Describe how to cut a boneless breast.

10. Describe how to cut an airline breast.

11. Cut a boneless breast and an airline breast.

12. Describe how to bone legs and thighs.

13. Bone a leg and a thigh.

14. Describe how to partially bone legs and thighs.

15. Partially bone a leg and a thigh.

16. Describe how to bone whole poultry.

17. Bone whole poultry.

FISH

A *fish* is any of a classification of animal that has fins, gills, and an internal skeleton made of bones or cartilage. Fish have edible flesh that contains protein, carbohydrates, fat, water, and trace amounts of vitamins and minerals. Unlike land animals, fish have delicate connective tissue. Because the connective tissue is very thin, fish flesh naturally flakes and comes apart easily. Fish are considered lean or fatty based on the amount of fat they contain. **See Figure 4-12.**

Production Tip

Avoiding the use of seafood that is endangered or being overharvested is key to sustainability. The Monterey Bay Aquarium's Seafood Watch program provides lists of seafood to avoid as well as substitution options.

Approximate Fat Content of Common Fish			
Lean Fish		Fatty Fish	
Species	Fat Content*	Species	Fat Content*
Cod	1 g	Farmed catfish	6 g
Flounder	1 g	Rainbow trout	4 g
Halibut	2 g	Coho salmon	6 g
Striped bass	2 g	Shad	10 g

* per 3 oz serving

Figure 4-12. Fish are considered lean or fatty based on the amount of fat they contain.

The fat content of a fish can affect both flavor and the cooking method required. Lean fish contain very little fat and are often prepared using moist-heat cooking methods. Pacific cod is an example of a lean fish. Fatty fish contain more fat and are rich in omega-3 fatty acids, vitamin A, and vitamin D. Fatty fish, such as salmon, are best prepared using dry-heat cooking methods. The fat content of a particular type of fish can vary slightly by season.

Fish are most often grouped by freshwater or saltwater habitat and by their external shape and structure. Fish are classified as roundfish, flatfish, or cartilaginous fish, based on their shape and structure.

U.S. Fish & Wildlife Service

Roundfish

A *roundfish* is any fish with a cylindrical body, an eye located on each side of the head, and a backbone that runs from head to tail in the center of the body. **See Figure 4-13.** Roundfish are the most common type of fish and are found in freshwater lakes and streams as well as in saltwater. Trout and salmon are examples of roundfish.

Roundfish Shape and Structure

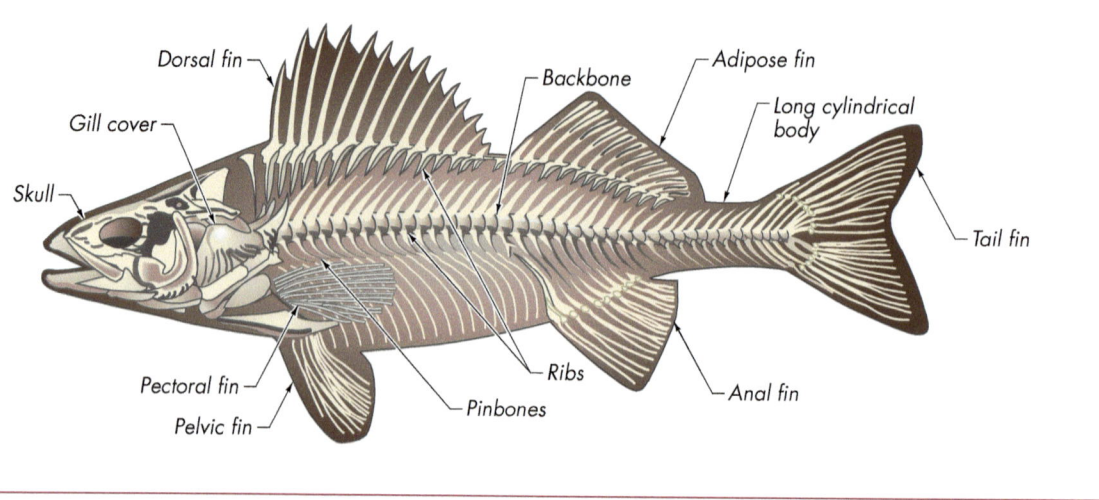

Figure 4-13. A roundfish is any fish with a cylindrical body, an eye located on each side of the head, and a backbone that runs from head to tail in the center of the body.

Flatfish

A *flatfish* is any thin, wide fish with both eyes located on one side of the head and a backbone that runs from head to tail through the lateral line of the body. **See Figure 4-14.** Flatfish are born round like other fish. However, as they grow, their eyes move to one side of the body so they can see all around when lying flat on the ocean or sea floor. Flatfish swim parallel to the surface of the water with both eyes facing toward the surface. The skin on the top side of a flatfish is typically dark and may change color to blend in with its environment. The bottom side of a flatfish is typically light in color. Flounder, halibut, and sole are examples of flatfish.

Cartilaginous Fish

A *cartilaginous fish* is any fish that has a skeleton composed of cartilage instead of bones. **See Figure 4-15.** Cartilaginous fish often have a smooth, tough outer skin without scales. Sharks, skates, and stingrays are examples of cartilaginous fish.

Flatfish Shape and Structure

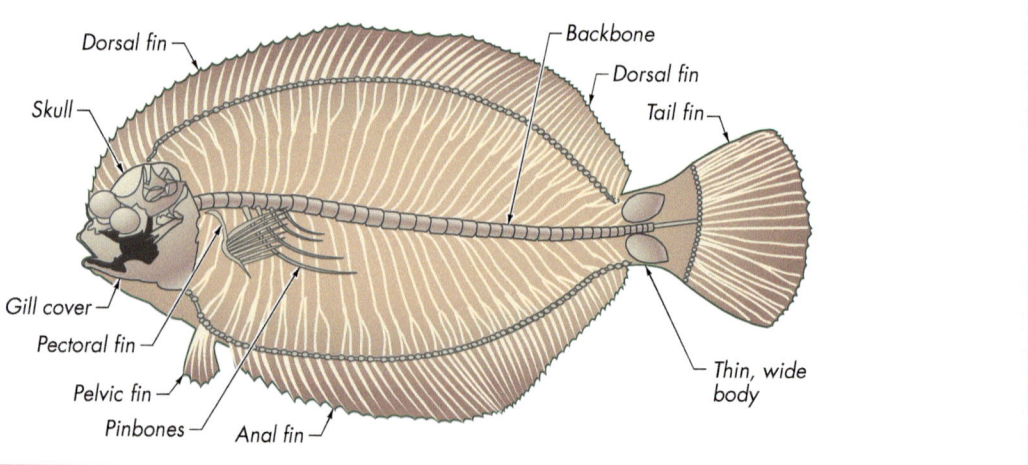

Figure 4-14. A flatfish is any thin, wide fish with both eyes located on one side of the head and a backbone that runs from head to tail through the lateral line of the body.

Cartilaginous Fish Structure

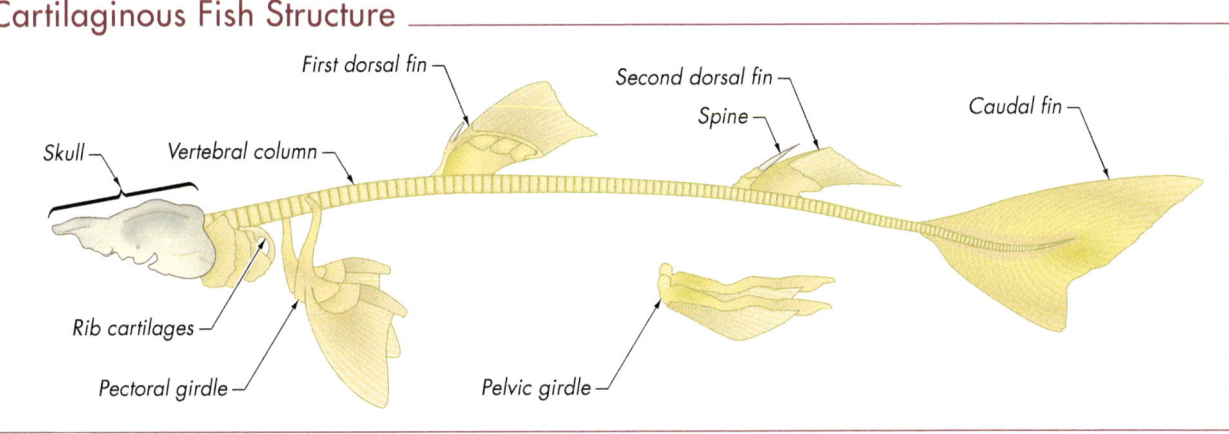

Figure 4-15. A cartilaginous fish is any fish that has a skeleton composed of cartilage instead of bones.

CHECKPOINT 4-4

1. Differentiate between lean and fatty fish.

2. Describe three classifications of fish based on external shape and structure.

MARKET FORMS OF FISH

Fish may be purchased fresh, frozen, or processed. The more that is done to a fish prior to delivery, the higher the cost per pound. Fish are commonly sold whole, drawn, dressed or pan-dressed, as steaks, as fillets, as loins, and as wheels. **See Figure 4-16.**

Market Forms of Fish

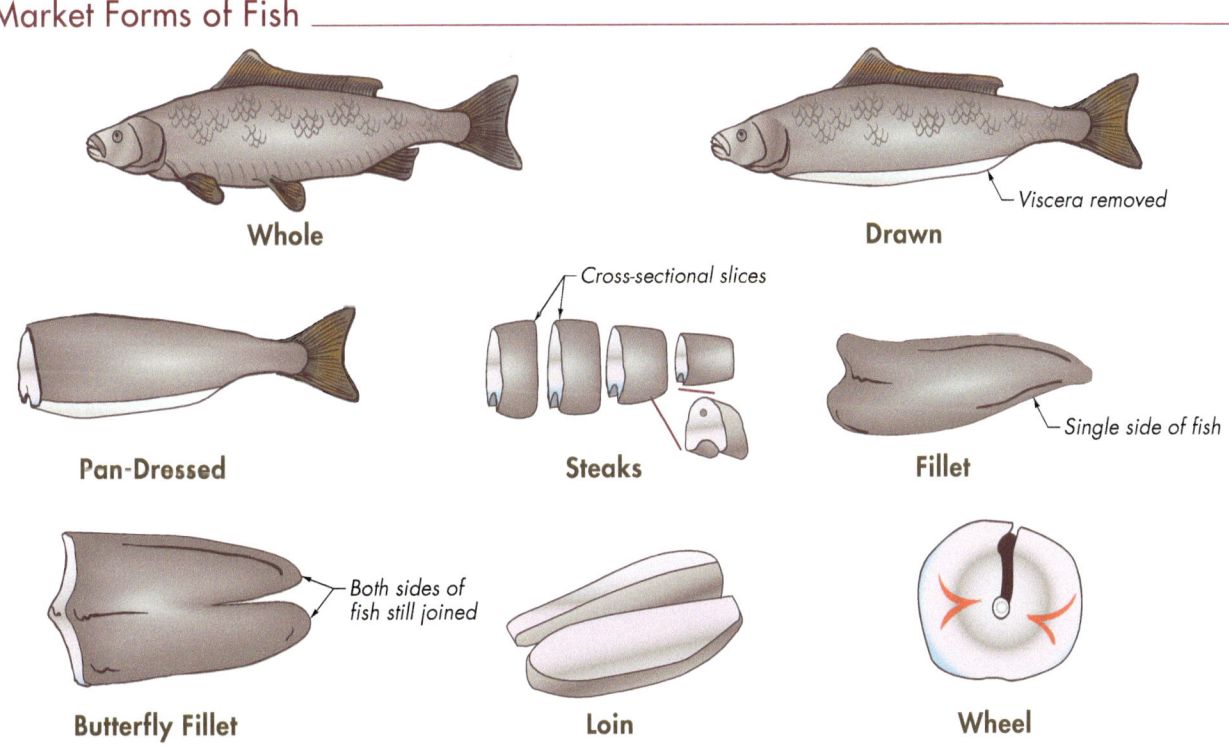

Figure 4-16. Fish are commonly sold whole, drawn, dressed or pan-dressed, as steaks, as fillets, as loins, and as wheels.

Whole Fish

A *whole fish* is the market form of a fish that is taken from the water and sold as is. Nothing has been done to process it. Whole fish have the shortest shelf life of any market form because all the internal organs (viscera) are still present. Fish purchased whole cost less per pound, require more preparation, and yield more waste than any other market form.

Drawn and Dressed Fish

A *drawn fish* is a fish that has had only the viscera removed. Drawn fish can be prepared whole. A *dressed fish* is a fish that has been scaled and has had the viscera, gills, and fins removed. A *pan-dressed fish,* also known as a headed and gutted (H&G) fish, is a dressed fish that has had its head removed.

Fish Steaks

A *fish steak* is a cross section of a dressed fish. Fish steaks from large fish are fabricated from a wheel. Steaks are ready to cook when purchased. Generally, the only bone present in a steak is a small section of the backbone. Steaks from very large fish, such as swordfish, are boneless. Tuna, salmon, swordfish, and sharks are commonly sold as steaks.

Fish Fillets

A *fish fillet* is the lengthwise piece of flesh cut away from the backbone. Roundfish have two fillets, one on each side. Flatfish have four fillets, two on each side. Fillets can be purchased with or without belly bones and pin bones. Fillets can also be purchased with or without skin. Fillets with the skin left on are sold scaled. A *butterflied fillet* is two single fillets from a dressed fish that are held together by the uncut back or belly of the fish.

Fish may also be sold as loins, wheels, roasts, or portions. A fish loin is cut lengthwise from either side of the backbone of large roundfish. A *wheel* is the round center cut of a large fish from which steaks are cut. A fish portion is cut from fillets into smaller pieces that are sold individually.

Frozen Fish

Using frozen fish allows a foodservice operation to serve a wider variety of fish year-round. Frozen fish cannot be purchased whole. For example, a large percentage of salmon is sold frozen. Frozen salmon is held at 32°F until it is flash-frozen. It is protected from dehydration by glazing. *Glazing* is the process of covering an item with water to form a protective coating of ice before the item is frozen. Proper glazing of frozen salmon results in a product that tastes like fresh salmon.

Processed Fish

Processed fish are available canned, smoked, salted, or pickled. Sardines, anchovies, and tuna are the most common varieties of canned fish. Canned fish should be checked for signs of damage or bulging before they are used. Damaged cans should be discarded.

Popular smoked and cured fish include cod, haddock, salmon, sturgeon, and herring. **See Figure 4-17.** Cod is often salted, whereas salmon and herring are often pickled.

Processed Fish

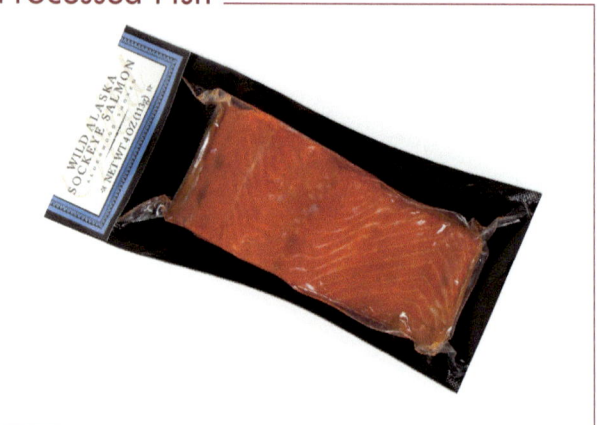

Figure 4-17. Salmon is often smoked.

Receiving and Storing Fresh Fish

Fresh fish spoils rapidly so it is essential to check the smell, appearance, and flesh upon receipt. **See Figure 4-18.** The smell of a fish is one of the easiest ways to determine freshness, but it can also be misleading. Fresh fish have a light, fishy smell. Strong fishy smells or ammonia odors are signs of deterioration. Fish that give off such smells should not be accepted.

The eyes and gills of fresh fish indicate age. The eyes should be round, slightly bulging, and clear. Cloudy or sunken eyes are signs of age and the fish should be rejected. The gills of a fresh fish should be bright red or reddish pink. If the gills are brown or missing, the fish should be rejected.

If fish has been cut into fillets, they should be moist and solid. Gently pressing on the fillet should not leave an indentation. The flesh should be firm and spring back when touched. A fillet should not be slimy, and the flakes of flesh should not separate when the fillet is bent slightly. The flesh should be of the proper color for the particular variety of fish. There also should not be any bruises or other signs of visible damage. If a fish shows any of these signs, it should be rejected.

Fabrication Station 17

Consideration for Purchasing Fresh Fish		
	Acceptable	*Unacceptable*
Smell	• Slight smell of seaweed or the ocean	• Strong fishy smell
	• Slight smell of fish	• Ammonia odor
Appearance	• Round, clear, bulging eyes	• Cloudy or sunken eyes
	• Bright-red or reddish-pink gills	• Brown gills or missing gills
	• Firmly attached scales	• Loose scales
	• Properly covered flesh	• Bruised, discolored, or damaged flesh
	• Moist and solid fillets	• Flesh of fillets separates when slightly bent
Touch	• Wet, slightly slippery exterior surface	• Slimy internal cavity
	• Smooth scales lying flat against body	• Rough scales
	• Moist, intact fins	• Dry fins
	• Firm flesh	• Mushy flesh

Figure 4-18. Fresh fish spoils rapidly so it is essential to check the smell, appearance, and flesh upon receipt.

The external surface of a fresh fish should be wet and a little slippery. Scales should be smooth and lie flat against the body of the fish. Scales falling off the fish are signs of age, and these fish should be rejected. Fins should be moist and intact. If the fins are dried out or removed, the fish may be older and the flesh may be starting to dry out. Fish with dry fins should be rejected.

Even if the fish being delivered is from an inspected facility, it may not have been held at proper temperatures since leaving the plant. Fresh fish should be packed in flaked ice to maintain a safe temperature. Fish should have an internal temperature of 41°F or lower to be accepted at delivery. Fish above 41°F should be rejected. Once accepted, fish must be stored at 41°F or below and used as quickly as possible to maintain quality and freshness.

Fresh fish can be stored a maximum of two days. If a fresh fish is not cooked upon receipt, it should be removed from the ice it was packaged in, placed in a self-draining container, covered with plastic wrap, and then covered with a think layer of crushed ice. **See Figure 4-19.** Then the container should be placed in the coldest part of the refrigerator. The plastic wrap helps prevent the flesh of the fish from developing freezer burn from the ice and from absorbing water as the ice melts. The ice helps hold the proper temperature and reduces deterioration. The ice should be

replaced daily. Fish must be stored away from other food in the refrigerator to prevent the odors from affecting the smell or taste of other foods.

Storing Fresh Fish

Figure 4-19. If a fresh fish is not cooked upon receipt, it should be removed from the ice it was packaged in, placed in a self-draining container, covered with plastic wrap, and then covered with a thin layer of crushed ice.

Receiving and Storing Frozen Fish

When receiving frozen fish, it should be received frozen solid. If the fish is even slightly thawed on the edges, it should be rejected. Fish that has been refrozen will have a poor texture. If there are dry-looking spots on the fish, it has slightly thawed and been refrozen and should be rejected. Signs of moisture on the outside of the box may also indicate that the product has thawed and been refrozen. Ice crystals are another sign that fish thawed or was not stored at or below 0°F. Frozen fish quickly deteriorate at temperatures above 0°F, so fish displaying ice crystals should be rejected.

Fish stored in the freezer must be wrapped properly with moistureproof wrapping to prevent freezer burn. Fatty fish should not be stored in the freezer for more than two months because they deteriorate quickly. Lean fish can be stored frozen for up to six months. Stock should be rotated so that the oldest fish is used first. Frozen fish must be thawed in the refrigerator as close as possible to the time the fish is needed. Once frozen fish has thawed, it must be used or discarded. Some seafood vendors offer a slacking or refreshing service. Slacking involves thawing the previously frozen seafood under refrigeration to ensure proper temperature control at the vendor's facility. The same quality indicators should be used as when receiving fresh fish.

Frozen fish can be individually quick-frozen (IQF), layer packed, cello-packed, or block frozen. *Individually quick-frozen (IQF)* is a designation for products preserved using a method in which each item is glazed with a thin layer of water and frozen individually. **See Figure 4-20.** IQF portions can be packaged together without sticking to other portions. This allows as much of an item to be used as needed without having to thaw an entire package. IQF products are packaged according to an average size, such as 2–3 oz fillets. IQF packaging makes it easy to remove the exact number of items needed.

Layer packs, also known as shatter packs, consist of high-quality, graded fish fillets layered on polyethylene sheets. The edges slightly overlap so that entire layers can be removed when desired.

Cello-packs contain ungraded fish fillets that are frozen in packets, typically one to three fillets per packet, wrapped in cellophane, frozen, and packaged six packets per box. Fillets packaged in this manner may be inconsistent in size and are relatively inexpensive.

Block-frozen fish is placed in a block-shaped form between two hollow stainless steel plates that have refrigerant flowing through them. The plates freeze the fish into a solid block within 2–4 hours.

Inspection and Grades of Fish

Fresh fish are not subject to a mandatory federal inspection. Instead, optional inspections are carried out by the National Marine Fisheries Service, which is part of the United States Department of Commerce (USDC). There are three types of optional inspections. These types of inspections include Type 1, Type 2, and Type 3.

A Type 1 inspection guarantees the fish or shellfish product is safe and wholesome for human consumption, is accurately labeled, has a good odor, and was processed in a sanitary, inspected facility. After being processed under a Type 1 inspection, a PUFI (processed under federal inspection) mark is affixed to the packing carton. **See Figure 4-21.** Type 1 inspection involves continual inspection of the fresh product from the time it arrives at the processing plant to the moment it is packaged for sale.

PUFI Marks

Figure 4-21. After being processed under a Type 1 inspection, a PUFI (processed under federal inspection) mark is affixed to the packing carton.

Individually Quick-Frozen (IQF) Fish

Drawn Trout

Orange Roughy Fillets

Mahi Mahi Steaks

Dressed Salmon

Halibut Steaks

Ahi Tuna Steaks

Czimer's Game & Seafoods, Inc.

Figure 4-20. Individually quick-frozen (IQF) is a designation for products preserved using a method in which each item is glazed with a thin layer of water and frozen individually.

A Type 2 inspection takes place in a warehouse or cold storage facility. The product is randomly inspected to ensure it meets the product specifications listed on a specification sheet.

A Type 3 inspection involves the examination of the fishing boats and processing plants to ensure that they are adhering to sanitation guidelines when handling and processing the product.

Only fish that is processed under a voluntary Type 1 inspection is eligible for grading. Fish may be graded A, B, or C. Because there are so many varieties of fish, the USDC only sets grade standards for the most common varieties. Grade A fish are of the best quality and do not have any visible defects. Grade B and Grade C are typically used for processed products.

Premium quality fish are used to make sashimi, which is masterfully presented small pieces of raw fish. Sashimi-grade fish must either be supercooled or kept alive until it is cut and served. *Supercooling* is the process of cooling an item below the freezing point without the item becoming solid or crystallizing. Sashimi-grade fish are typically supercooled to –76°F in order to maintain their texture and quality. The extremely low temperature also kills any parasites that could cause foodborne illnesses.

Production Tip

The practice of treating sashimi-grade fish with carbon monoxide to keep the flesh red for a longer time in storage can make spoiled fish appear fresh. It is best to purchase supercooled or live fresh fish for use as sashimi.

CHECKPOINT 4-5

1. Describe the various market forms of fish.

2. Explain how fresh fish are received and stored.

3. Explain how frozen fish are received and stored.

4. Name the government organization in charge of voluntary fish inspections.

5. Describe what happens during a Type 1 inspection.

6. Explain the importance of supercooling sashimi-grade fish if not using live fish.

FABRICATING FISH

Fish may be purchased whole and fabricated in-house or may be purchased in a portion-controlled or processed form. Knowing and using proper fabrication techniques for fish is an important skill to have in the professional kitchen.

Scaling Fish

A fish scaler is used to remove the scales from fish prior to preparation. **See Figure 4-22.** However, care should be taken to ensure the flesh is not damaged as the scales are removed.

Fish Scalers

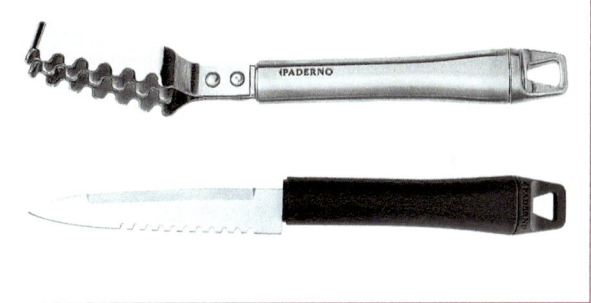

Paderno World Cuisine

Figure 4-22. A variety of fish scalers can be used to remove the scales from fish.

Procedure for Scaling Fish

1. In a sink, firmly hold the tail of the fish with the guiding hand. Use a fish scaler to scrape against the scales down toward the head. *Note:* Use caution not to damage the flesh of the fish by applying too much pressure.

2. When finished scaling one side, turn the fish over and repeat the process.

3. Rinse the fish under cold running water to remove any loose scales.

A fish steak is a popular cut for many large roundfish varieties. For example, tuna is often purchased in loins. The loins are commonly cut into steaks. Belly loins commonly have a higher fat content than back loins.

A *blood line* is a dark-red, almost black muscle that runs along the lateral line of a fish. The blood line should be removed prior to preparation. There may also be a vein running along the loin that should be removed because it has a strong, undesirable flavor.

Fabricating Fish

Procedure for Cutting Roundfish into Steaks

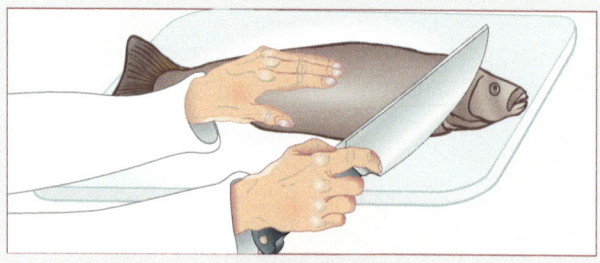

1. Use a boning knife to remove the scales, fins, viscera, head, and tail from the fish.

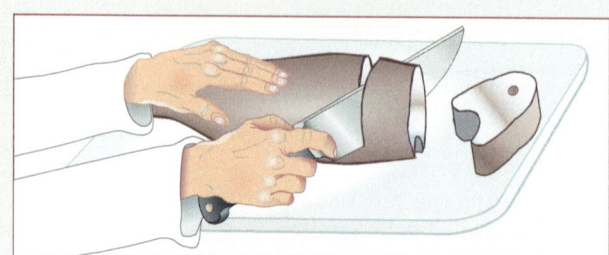

2. Slice the fish horizontally into equal sized pieces.

Filleting Roundfish

A fillet is the most common serving portion from a fish. It is important to remove all belly bones and pin bones from a fillet before cooking.

Skinning Fillets

A fillet can be prepared with or without the skin intact. A particular presentation or dish may require removing the skin from a fillet.

Procedure for Filleting Roundfish

1. Use a boning knife to make a cut about ½ inch behind the gills and down to, but not through, the backbone.

2. Make a second cut along the backbone from just behind the head all the way to the tail. Do not cut through the backbone.

3. Starting at the tail, carefully slice toward the head to cut the flesh away from the backbone.

4. Carefully lift the fillet and cut away any rib bones that are still attached to the fillet. Trim any belly fat from the fillet.

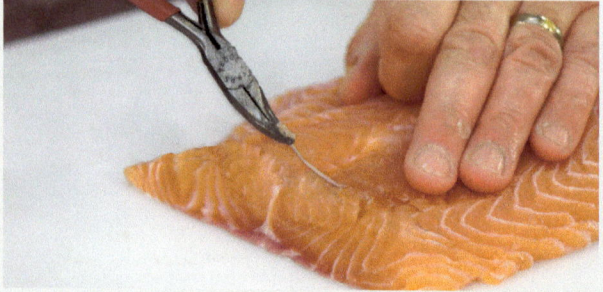

5. Run fingers gently along the surface of the fillet to raise the ends of any pinbones that may remain. Use needle-nose pliers to remove the pinbones.

6. Turn the fish over and repeat the entire process on the other side.

Filleting Roundfish
Media Clip

Procedure for Skinning Fillets

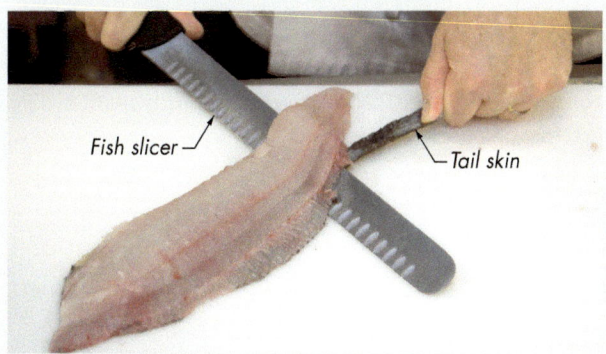

1. Place the fillet skin-side down. Starting at the tail, use a fish slicer or boning knife to carefully cut down through the flesh to the skin while using the guiding hand to firmly hold the tail skin.

2. Angle the edge of the blade slightly downward and cut between the skin and the flesh while moving the blade toward the head end.

Skinned fish fillet.

Filleting Flatfish

The backbone of a flatfish runs through the midline of the fish. As a result, flatfish yield four fillets, two on each side of the fish.

Production Tip

Using a pair of kitchen shears to remove the fins from both sides of a flatfish will make filleting the fish easier.

Procedure for Filleting Flatfish

1. Use a boning knife to cut along the backbone from the head to the tail.

2. Insert the tip of the blade near the head of the fish and carefully slice the flesh away from the bones on one side of the backbone.

3. Remove the second fillet using the same process.

4. Turn the fish over and repeat steps 1–3 to remove the other two fillets.

Filleting Flatfish
Media Clip

CHECKPOINT 4-6

1. Explain how to scale a fish.

2. Scale a fish.

3. Explain how to cut a roundfish into steaks.

4. Cut a roundfish into steaks.

5. Explain how to fillet a roundfish.

6. Fillet a roundfish.

7. Explain how to skin a roundfish fillet.

8. Skin a roundfish fillet.

9. Explain how to fillet a flatfish.

10. Fillet a flatfish.

SHELLFISH

Shellfish is the classification of aquatic invertebrates that may or may not have a hard, external shell. An external shell functions as a skeleton and is called an exoskeleton. Shellfish are commonly categorized as crustaceans or mollusks. **See Figure 4-23.**

Crustaceans

A *crustacean* is a shellfish that has a hard, segmented shell that protects soft flesh and does not have an internal bone structure. Crustaceans live in both freshwater and saltwater. Unlike fish, crustaceans can live out of water for a few days if they are kept moist. Crustaceans include shrimp, prawns, lobsters, crayfish, and crabs.

Shrimp and Prawns. Shrimp and prawns are different species of crustaceans that both have a tender white flesh. However, the terms shrimp and prawns are often used interchangeably. In the United Kingdom and Australia, the term prawn is used for both shrimp and prawns. In North America, larger shrimp are commonly marketed as prawns. **See Figure 4-24.**

Shrimp and Prawns

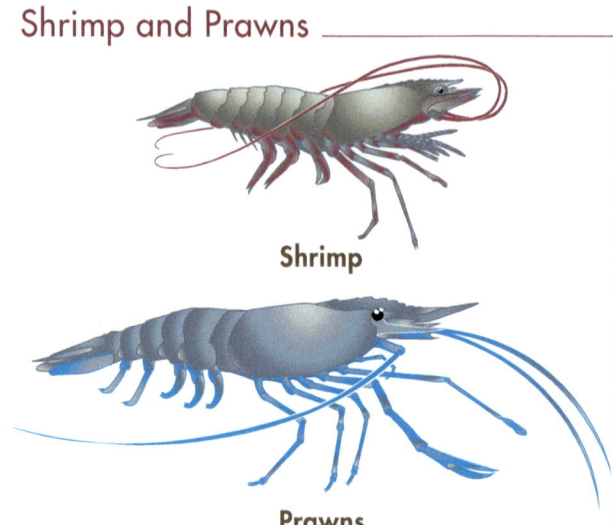

Shrimp

Prawns

Figure 4-24. In North America, larger shrimp are commonly marketed as prawns.

Four common types of shrimp used in the professional kitchen are white, brown, pink, and black tiger. Although these four types of shrimp vary in color when caught, they differ very little in appearance when cooked. They also have similar flavor and nutritional value. Grilling, broiling, sautéing, and frying are popular methods of preparing shrimp. Cooked shrimp also can be chilled and served cold as appetizers, salads, or entrées.

Shellfish

Agricultural Research Service, USDA

Crustaceans

Florida Department of Agriculture and Consumer Services, Bureau of Seafood and Aquaculture Marketing

Mollusks with Shells (Univalves)

National Oceanic and Atmospheric Administration

Mollusks without Shells (Cephalopods)

Figure 4-23. Shellfish are commonly categorized as crustaceans or mollusks.

Lobsters. A *lobster* is a saltwater crustacean with a brown to bluish-black external shell and two large claws. **See Figure 4-25.**

Lobsters are most often steamed or poached. The two primary types of lobster are clawed lobsters and spiny lobsters. The two types are similar in size but vary in color. Clawed lobsters contain more meat than spiny lobsters. Clawed lobsters are sold live whereas spiny lobsters are harvested for their tails. Common varieties of lobster include Maine lobsters, spiny lobsters, and langoustines.

- A *Maine lobster* is a large clawed lobster with a dark bluish-green shell, two heavy claws, and eight slender legs. Maine lobsters are harvested from the cold North Atlantic waters.
- A *spiny lobster,* also known as a rock lobster, is a lobster that has spines covering its body and five slender legs on each side. Spiny lobsters have smaller claws than Maine lobsters. The tail section of a spiny lobster is typically eaten. Spiny lobsters are typically harvested from warm waters.
- A *langoustine,* also known as a Norway lobster or a Dublin Bay prawn, is a small lobster that resembles a very large shrimp, except it has very long front arms with long, thin claws. Only the tail section of the langoustine is eaten. Langoustines are harvested from both cold and warm waters such as the northeastern Atlantic Ocean, the North Sea, and the Mediterranean Sea.

Lobsters that are sold live must be kept alive until cooking. A *sleeper* is a lobster that is dying. Sleepers are sold at a reduced price and must be cooked immediately. Cooking and eating dead lobsters can be harmful.

Almost all lobster flesh is edible except for a small section near the eye area. Most of the flesh is found in the tail and the claws. The soft green substance found in the body cavity of a lobster is the liver and pancreas, which are collectively called the tomalley. The tomalley is considered a delicacy. It may be eaten alone or used to thicken sauces. Lobster paste, also known as lobster pâté, is a mixture of tomalley and lobster roe (coral).

Crayfish. A *crayfish,* also known as crawfish or crawdad, is a freshwater crustacean that resembles a tiny lobster. **See Figure 4-26.** Crayfish are dark brown to black in color and can range from 3–7 inches in length. Most of the crayfish harvested come from Louisiana and the Pacific Northwest. Crayfish have a flavor similar to that of shrimp, but a slightly tougher texture. They are commonly used in Creole, Cajun, and French cuisine.

Crayfish

Figure 4-26. A crayfish, also known as crawfish or crawdad, is a freshwater crustacean that resembles a tiny lobster.

Crabs. Crabs are a popular variety of shellfish with tender, sweet-tasting flesh that can be used in many menu items. Crabmeat is available fresh, frozen, and canned. When using fresh crabs, only live crabs should be used. Fresh crabs that have died should be discarded.

Anatomy of a Lobster

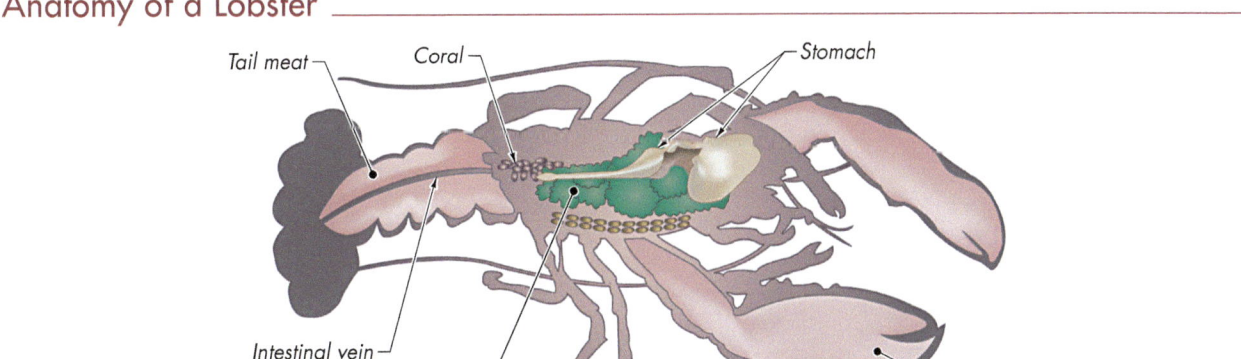

Tail meat — Coral — Stomach
Intestinal vein — Tomalley — Claw meat

Figure 4-25. A lobster is a saltwater crustacean with a brown to bluish-black external shell and two large claws.

Most edible crabs can be distinguished from inedible varieties by counting the pairs of legs. Most edible crabs have five pairs of legs, four for walking and one pair that serve as arms. Most inedible crabs have only four pairs of legs. King crabs are an exception, as they have only four pairs of legs but are still edible. King crabs, blue crabs, Dungeness crabs, snow crabs, and stone crabs are frequently used in the professional kitchen. **See Figure 4-27.**

- *King crab* is the largest-sized variety of crab, typically weighing between 6–20 lb and measuring as much as 10 feet from the tip of one leg to the tip of the opposite leg. King crabs are one of the few edible crabs that have four pairs of legs. The raw flesh has a pinkish tinge that becomes snowy white when cooked. King crab is available in the shell as cooked frozen legs, as legs and claws, as leg and body flesh, or as shredded flesh. Canned or frozen king crabs are also available shucked or pulled (removed from the shell). King crab is not typically sold live because the legs are removed at catch and flash-frozen to preserve their quality.

- *Blue crab* is a North American crab with blue claws and a dark blue-green, oval shell. Blue crabs measure approximately 5 inches across and weigh approximately 5 oz. Hard-shell blue crab is sold live, cooked and frozen, or cooked, pickled, and canned. A *soft-shell crab* is a blue crab that has been harvested within 6 hours of molting, or shedding its shell in order to grow a larger shell. Molting season is from mid-May to the beginning of September. Soft-shell crab is only sold live during molting season. Soft-shell crabs must be handled with special care to ensure that they arrive at their destination alive. They must be kept alive until they are cooked or cleaned and immediately quick-frozen.

- *Dungeness crab* is a Pacific crab with a sweet-tasting flesh. Approximately 25% of the body weight of a Dungeness crab is flesh, which is the highest percentage found in any variety of crab. Most of the edible flesh in a Dungeness crab is body flesh, not leg flesh. Dungeness crabs are larger than blue crabs, weighing approximately 1¾–4 lb each. Dungeness crab is commonly available frozen or canned but can be purchased live.

- A *snow crab* is a crab that is similar to a king crab but is smaller and available in greater supply. Also known as a spider crab, snow crab is typically sold as cooked, frozen leg clusters. A snow crab has only a fraction of the flesh that a king crab has. Frozen leg clusters are commonly steamed or poached just enough to warm the flesh because they are already fully cooked. Snow crab can be steamed and served whole or served cold in a crab cocktail.

Crab Varieties

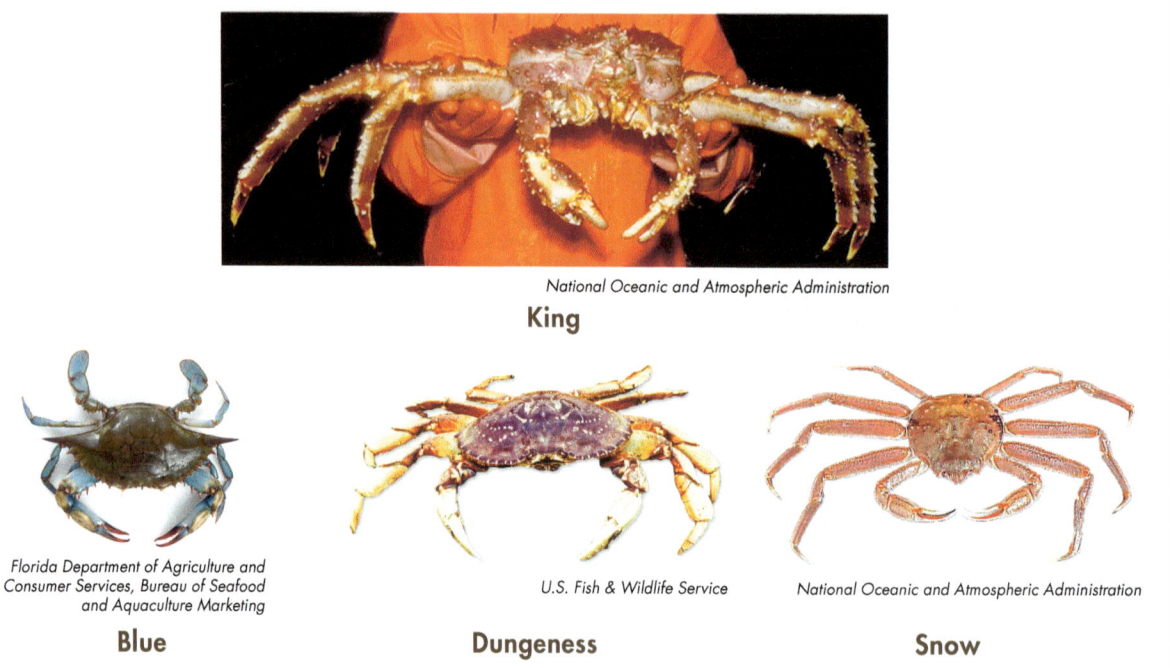

National Oceanic and Atmospheric Administration
King

Florida Department of Agriculture and Consumer Services, Bureau of Seafood and Aquaculture Marketing
Blue

U.S. Fish & Wildlife Service
Dungeness

National Oceanic and Atmospheric Administration
Snow

Figure 4-27. Crab varieties include king crab, blue crab, Dungeness crab, snow crab.

- A *stone crab* is an Atlantic crab with a brownish-red shell and large claws of unequal size. Unlike other varieties of crab, when a stone crab is caught one claw is removed and the crab is placed back in the water. The claws are the only parts of the stone crab that are harvested. In approximately 12–24 months, the stone crab generates a new claw in place of the one that was removed. Each time a claw is removed it grows back slightly larger than before. Stone crab claws range from 2–5 oz and are typically served cold with most of the shell removed. They can also be served hot. Stone crabs have a similar taste and texture to lobster.

Mollusks

A *mollusk* is a shellfish with a soft, nonsegmented body. Some mollusks, such as clams, have a hard, external shell. Other types of mollusks, such as squid, do not have an external shell. Like crustaceans, mollusks do not have an internal skeleton. Mollusks should always be alive prior to cooking, unless they are frozen or canned. If a mollusk dies prior to cooking, it should be discarded. The three classifications that edible mollusks are divided into are univalves, bivalves, and cephalopods.

Univalves. A *univalve*, also known as a gastropod, is a mollusk that has a single solid shell and a single foot. A univalve uses its foot to move along the surface of underwater structures or on land. Univalve varieties include abalone and conch. **See Figure 4-28.**

- An *abalone* is a univalve contained in a brown, bowl-shaped shell with an iridescent, multicolored interior. The California abalone variety is considered the highest in quality, but all varieties of abalone have a sweet, slightly salty flesh. The texture of an abalone is tender like lobster, but has a creamier taste. It is very tough when overcooked. Aquafarmed abalones are raised in saltwater pens. Abalones are also imported from the coastal waters of Japan, New Zealand, and Mexico and are sold in canned and frozen form.

- A *conch* is a univalve that has a pinkish-orange shell with an interior that resembles a large snail. Conchs are found in the warm waters off the Florida Keys and throughout the Caribbean. Conch flesh is rubbery and is typically sliced thin and tenderized before it is cooked. It has a sweet flavor similar to that of a clam. Fresh conch is often eaten raw with lime juice and hot sauce.

Bivalve. A *bivalve* is a mollusk with a top shell and a bottom shell connected by a central hinge that can close for protection. Live bivalves should close tightly when gently tapped. Bivalves that are broken or do not close when tapped are dead and should be discarded. A bivalve that is noticeably heavier than others is probably dead and should be discarded. Common types of bivalves include clams, cockles, mussels, oysters, and scallops.

Nutrition Note

Mollusks are a good source of protein, vitamin B12, vitamin C, iron, zinc, and copper. Mollusks are also very high in cholesterol.

Univalves

National Oceanic and Atmospheric Administration

Abalone

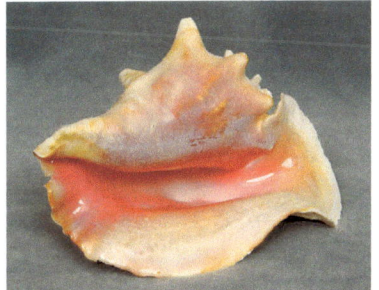

U.S. Fish & Wildlife Service

Conch

Figure 4-28. Univalve varieties include abalone and conch.

Clams. A *clam* is a bivalve found in both freshwater and saltwater. Clams can be fried or steamed with excellent results and are the key ingredient in clam chowders. There are many varieties of ocean clams found in different oceans. **See Figure 4-29.**

Ocean Clams

United States Department of Agriculture
Littlenecks

Plitt Seafood
Manila

Figure 4-29. There are many varieties of ocean clams.

The Atlantic ocean produces soft-shell, hard-shell, and surf clams. A *soft-shell clam,* also known as a long-neck or steamer clam, is an Atlantic clam with a thin, brittle shell that breaks easily. Soft-shell clams have a protruding siphon that prevents the shell from closing completely, which causes them to dry out more quickly than hard-shell clams once removed from the water. A *siphon* is a tubular organ that is used to draw in or eject fluids. Soft-shell clams also have a tendency to be gritty due to excess sand settling inside the shell. They need to be soaked in a solution of salted water and cornmeal to remove the sand. Soft-shell clams have a tender, sweet-tasting flesh and are commonly steamed or fried.

A *hard-shell clam,* also known as a quahog, is an Atlantic clam with a blue-grey shell that contains a chewy flesh. These clams are rarely sold by the name hard-shell clam or quahog. Instead, they are sold by classifications that indicate their size. These classifications include littleneck, cherrystone, topneck, and chowder clams.

- Littlenecks are 1–2 inches in size. These bite-size clams are served raw on the half shell or steamed.
- Cherrystones are 2–3 inches in size. They are commonly steamed or served raw on the half shell.
- Topnecks are 3 inches in size and are often stuffed and baked.
- Chowder clams are larger and typically cut into strips or minced for making soups and chowders.

A *surf clam* is a large species of Atlantic clam that can grow to 8 inches in size. Half of a shucked surf clam is the siphon, which is typically cut into strips and then breaded and fried. The other half of the shucked surf clam is the adductor muscle. The *adductor muscle* is a muscle that opens and closes the shell of a bivalve. It is commonly chopped or ground for use in chowders. Surf clams have a sweet, mild flavor.

Pacific clams have a tougher texture than Atlantic clams. Pacific clams include Manila clams, razor clams, butter clams, and geoducks.

Manila clams are the most common variety of Pacific clam. The shell is covered with slight ridges from the lip to the hinge. Manila clams can be steamed or served raw and have a sweet and salty flavor.

Razor clams are named for their narrow, oval-shaped shell. Razor clams have a sweet flavor. Butter clams have a mild, buttery flavor.

A *geoduck* is a very large Pacific clam with a meaty siphon that protrudes from its shell. Geoducks can weigh up to 3 lb. The siphon of a geoduck is typically split in half lengthwise and then cut into very thin slices and served carpaccio-style. It can also be quickly sautéed, steamed, or poached. Geoducks become very tough if overcooked.

Cockles. A *cockle* is a 1 inch wide bivalve with a shell that has deep ridges. **See Figure 4-30.** The small, fleshy cockles are often served shucked with a squeeze of lemon juice. Cockles are often used to make paella and Southeast Asian and Thai dishes.

Cockles

Fortune Fish Company

Figure 4-30. A cockle is a 1 inch wide bivalve with a shell that has deep ridges.

Mussels. A *mussel* is a freshwater or saltwater bivalve with whisker-like threads that extend outside the shell to allow the animal to attach to items for protection. These threads are referred to as a "beard." In the wild, mussels are commonly found attached by their beards to rocks. Aquafarmed mussels attach their beards to ropes. Common types of mussels include blue mussels and greenlip mussels.

- Blue mussels are the most common variety of edible mussel. **See Figure 4-31.** Aquafarmed blue mussels have a thinner, blue-black shell, while wild-caught blue mussels have a thicker, silver-blue shell. Blue mussels have tender, sweet flesh that is bright orange in color. They vary in size, but typically there are 10–20 mussels per pound. Blue mussels can be steamed and served either hot or cold or shucked and then sautéed or simmered in a sauce.

Blue Mussels

New Zealand Greenshell™ Mussels

Figure 4-31. Blue mussels are the most common variety of edible mussel.

- Greenlip mussels have a distinctive green-edged shell and are larger than blue mussels. They average 8–12 mussels per pound. They have a sweet, plump, and tender flesh. Greenlip mussels are often steamed with white wine, lemon, and herbs or cooked and served cold with cocktail sauce and lemon wedges.

Oysters. An *oyster* is a saltwater bivalve with a very rough shell that is coated with calcium deposits. Although oysters are not found in freshwater, they can be found in brackish water (a mixture of freshwater and saltwater). Most oysters are cultivated from oyster beds that require care and attention if they are to continue to produce. Common varieties include Atlantic oysters, Pacific oysters, and European oysters. The variety of oyster and where it comes from impacts the flavor profile.

- An *Atlantic oyster,* also known as an Eastern oyster, is a variety of oyster that has a fairly flat shell and a distinctive, salty-flavored, plump, and tender flesh. **See Figure 4-32.** Atlantic oysters account for roughly 70% of all oyster production. Common varieties of Atlantic oysters include Blue Point oysters, Chesapeake Bay oysters, Long Island oysters, and Prince Edward Island (PEI) oysters. PEI is known for high-quality, fresh seafood that can be delivered to the customer within hours of harvest.

Atlantic Oysters

Florida Department of Agriculture and Consumer Services, Bureau of Seafood and Aquaculture Marketing

Figure 4-32. Atlantic oysters have a fairly flat shell and a distinctive, salty-flavored, plump, and tender flesh.

- A *Pacific oyster,* also known as a Japanese oyster, is a variety of large oyster that has fragile, curvy shells and a briny, sweet, and mild-tasting flesh. The plump, moist flesh is silver, gold, or white in color. Common varieties include Olympia, Penn Cove, and Kumamoto oysters.
- A *European oyster* is a variety of oyster with a relatively flat, cup-shaped shell and salty-sweet flavored, creamy-textured flesh. European oysters are commonly served on the half shell. Belon oysters are a rare variety of European oysters.

Oysters are available year-round, but most are at peak quality from September to April. Oysters may be served raw, roasted, baked, breaded and fried, or poached. The key to preparing oysters is to apply just enough heat to heat them through, leaving them plump and tender. A common preparation is to top an oyster with a savory filling. Raw oysters need to be held on ice during service to keep them within a safe temperature range for consumption.

Scallops. A *scallop* is a bivalve with a fan-shaped shell and a cream-colored adductor muscle with a sweet, delicate flavor. The well-developed adductor muscle is the lean and juicy edible portion of the scallop. The rest of the scallop body is made up of white or red roe called coral, which is often removed prior to sale. Color is the best way to judge the quality of scallops. High-quality scallops have a creamy, almost translucent color. If scallops are white, they have been packed wet and the flavor and texture have been impaired. Rose-colored scallops indicate the scallops were in the process of producing roe. If scallops are a brownish color, they are old and should not be used.

Some chefs use scallop shells as serving dishes when featuring seafood appetizers or entrées. Bay scallops and sea scallops are different varieties of scallops that vary in size and flavor. **See Figure 4-33.**

Scallop Varieties

Bay

Sea

Figure 4-33. Bay scallops and sea scallops are different varieties of scallops.

A *bay scallop* is a fairly small scallop harvested from shallow saltwater. Bay scallops average 100 scallops per pound. Bay scallops are typically cleaned prior to sale and packed wet (soaked in a preservative that whitens the scallop and helps prevent spoiling), dry (untreated, without any preservatives), or individually quick-frozen. The most common preservative used with scallops is sodium tripolyphosphate, also known as STP. Bay scallops are commonly sautéed in butter, battered and fried, or marinated in lemon juice and served seviche-style.

A *sea scallop* is a large scallop with a coarse texture that is harvested from deep saltwater. Sea scallops are typically 2–5 times larger than bay scallops and average 30 scallops per pound. Sea scallops have a sweet, somewhat briny taste.

Fortune Fish Company

Cephalopods

A *cephalopod* is any of a variety of mollusks that do not have an external shell. Some cephalopods have an internal bone called a cuttlebone. Cephalopods have keen vision, a highly-developed nervous system, and a birdlike beak that is used to crack the shells of shrimp, crabs, lobsters, and other prey. Squid, octopuses, and cuttlefish are cephalopods. **See Figure 4-34.**

Squid. A *squid* is a translucent, head-footed cephalopod that has two tentacles, eight sucker-equipped arms, two lateral fins, and a flat, internal cuttlebone. The arms are attached near the eyes at the bottom of the head. The two tentacles are longer than the arms. Squid can change their skin color at will and expel a dark cloud of ink to confuse their prey. The ink is often used to color and flavor grain and pasta dishes. Most whole squid are packed 8–10 per pound. Larger squid are cut into steaks and sold frozen. Squid is often called by its Italian name, calamari. Squid is commonly sautéed, breaded and fried, or used to make stews. The flesh becomes tough if overcooked.

Octopuses. An *octopus* is a gray cephalopod with eight sucker-equipped arms, a birdlike beak, well-developed vision, and no internal or external shell. Octopuses can be small or quite large. However, they do not grow as large as giant squid. Octopuses are usually sold whole by the pound and are available fresh and frozen. When cooked, the gray skin turns deep purple or reddish-purple. Octopus flesh is white, firm, and sweet. The flesh becomes tough if overcooked. Cold, cooked octopus is used to make pulpo salad.

Cuttlefish. A *cuttlefish* is a translucent cephalopod with two tentacles, eight sucker-equipped arms, a hard internal cuttlebone, and large eyes at the base of its head. They can change their color at will. Cuttlefish expel a dark-brown cloud of ink to confuse their prey. The tentacles, arms, and ink are the only parts of a cuttlefish that are eaten. The ink is commonly used to make cuttlefish risotto, which is often dark in color. The tentacles and arms are typically cut into rings and sautéed, breaded and fried, or cooked in soups or stews. The flesh becomes tough if overcooked.

Cephalodpods

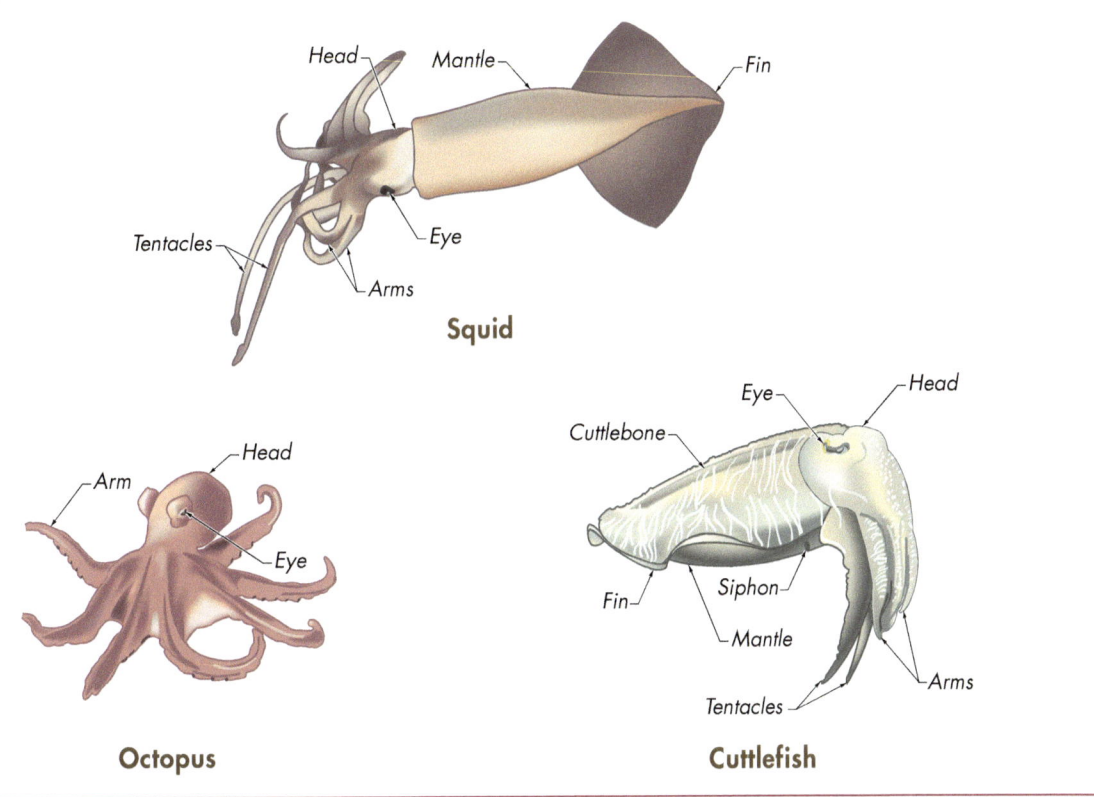

Figure 4-34. Squid, octopuses, and cuttlefish are cephalopods.

CHECKPOINT 4-7

1. Describe four types of crustaceans.

2. Describe two types of univalves.

3. Describe five types of bivalves.

4. Describe three types of cephalopods.

MARKET FORMS OF SHELLFISH

Each market form of shellfish has advantages and disadvantages in terms of cost, convenience, and labor. Shellfish are available live, shucked, or frozen. Shucked and cooked shellfish are also available canned.

Live Shellfish

Lobsters are usually purchased live and come in various sizes. On the coast both hard-shell and soft-shell crabs are sold live. Away from the coast only soft-shell crabs are available live.

Live clams, oysters, and scallops can be purchased by the gallon or the pound. They are usually packed by count. A higher count indicates a smaller size. If a container of oysters is marked 350 per gallon and another container is marked 225 per gallon, the 350 count container holds more oysters that are smaller. Live oysters and clams have tightly closed shells. If a shell is open or does not close fully when lightly tapped, the animal is dead and must be discarded. Oyster and clam shells that are unusually heavy probably contain mud and should also be discarded.

Shucked Shellfish

Clams, oysters, and scallops are often sold with the shell removed and are available both fresh and frozen. Shrimp may also be purchased cooked and canned. Cooked shrimp may be purchased by the pound either in the shell or peeled. Crabmeat can be cooked (fresh or frozen) and canned. Crabmeat also may be purchased fresh or frozen in the shell in a variety of cooked forms. **See Figure 4-35.** Cooked crabmeat should be kept packed in ice and refrigerated until it is used.

Forms of Cooked Crab	
Forms	*Descriptions*
Jumbo lump	Large pieces of white body flesh attached to the swimming legs
Back fin, lump, or special	Small pieces of white flesh from the body
Claws and fingers	Brownish flesh from the claws and legs
Cocktail claws	Claw flesh with a protion of the shell remaining as a handle for eating

Figure 4-35. Crabmeat also may be purchased fresh or frozen in the shell in a variety of cooked forms.

Imitation crabmeat, known as surimi, is also available. *Surimi* is a fish product made from a mixture of fish and/or shellfish and other ingredients. Surimi looks, cooks, and tastes like crabmeat. **See Figure 4-36.** It is high in protein and low in calories, sodium, fat, and cholesterol. Surimi is precooked and frozen. It can be purchased as legs, chunk meat, or flake meat.

Surimi

Harbor Seafood, Inc.

Figure 4-36. Surimi looks, cooks, and tastes like crabmeat.

Frozen Shellfish

Prepared lobster tails may be purchased frozen. The head and thorax (body) of shrimp and prawns are typically removed, and the tails are frozen while still at sea to ensure maximum freshness. Frozen shellfish is commonly packaged by count. **See Figure 4-37.** For example, shrimp packed and labeled 21/25 count indicates that there is an average

of 21–25 shrimp per 1 lb package. Frozen shrimp can be purchased green (uncooked), peeled or unpeeled, cooked and peeled, or peeled, deveined, and breaded. Frozen shrimp are usually sold in 5 lb blocks.

Packaging by Count

Sizing Shrimp	
Count	*Size*
25 or fewer per lb	Jumbo
25 to 30 per lb	Large
30 to 42 per lb	Medium
42 or more per lb	Small

Frozen Shrimp

Fortune Fish Company

King Crab Leg Clusters

Figure 4-37. Frozen shellfish is commonly packaged by count.

Receiving and Storing Shellfish

To be acceptable upon delivery, live shellfish should be delivered with an air temperature of 45°F and shucked shellfish must have an internal temperature of 45°F or below. Once accepted, it is imperative to store shellfish between 30°F and 34°F or below and to use it as quickly as possible to maintain quality. If the shellfish will not be used within two days of purchase, it should be wrapped in moistureproof freezer paper and foil, to protect it from air and moisture, and then frozen. Shrimp, scallops, and crabmeat are often block frozen.

Live Shellfish. Lobsters, crabs, clams, oysters, and mussels are often purchased live in the shell. Live crustaceans should be covered with wet seaweed or damp newspaper to keep them from drying out. Crustaceans may also be stored in a saltwater tank. In the absence of a tank, live crustaceans live two to four days. When received, the lobsters must be carefully inspected. Live lobsters have a tightly curled tail. Live lobsters and crabs should show leg movement when received and must be kept alive until they are cooked. If a shellfish is dead, it should be rejected.

Production Tip

To pick up a lobster, grab it by the back just above where the tail connects to the body. Stay clear of the claws to avoid injury, even when the claws are bound with rubber bands.

Fresh shellfish are not subject to a mandatory federal inspection. However, all mollusks must be delivered with a shellstock tag. Shellstock tags are waterproof, tear-resistant tags attached to containers of mollusks. **See Figure 4-38.** The tags must remain attached to the bag of shellfish until the bag is empty. Shellstock tags list the dealer's contact information and identification number, the original harvester (if different than the dealer), the harvest date and general area, the type and quantity of shellfish, a 90 day retention notice, and a consumer advisory. Shellstock tags are kept on file for 90 days after the mollusks are harvested in case of foodborne illness. Mollusks delivered without shellstock tags should be rejected.

Shellstock Tags

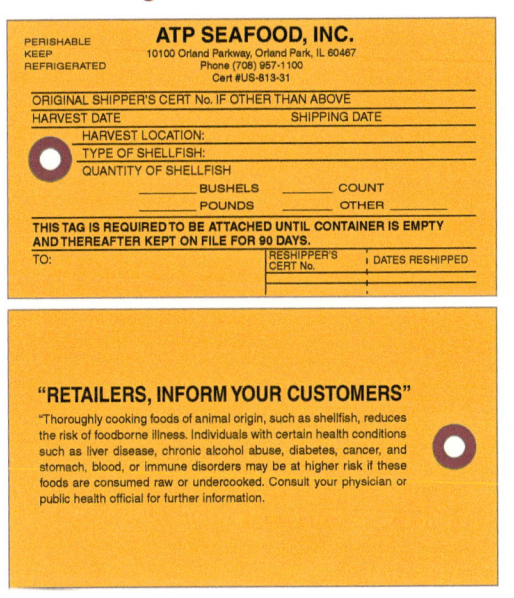

Figure 4-38. Shellstock tags are waterproof, tear-resistant tags attached to containers of mollusks.

Live bivalves should be refrigerated in the original box or netted bag and placed in a pan to prevent drips from contaminating other foods. Live mollusks should not be stored in a sealed container or plastic bag because they will die from a lack of oxygen. Under ideal conditions, fresh mollusks can live in refrigerated storage for up to a week. **See Figure 4-39.** Live oysters and clams have tightly closed shells. If a shell is open and does not close when handled, the oyster or clam is dead and should be discarded.

Storing Live Mollusks

Figure 4-39. Under ideal conditions, fresh mollusks can live in refrigerated storage for up to a week.

Shucked oysters are packed in metal containers and must be kept refrigerated and packed in ice at all times to prevent spoilage. If handled in the proper manner, oysters remain fresh for up to a week.

Frozen Shellfish. When receiving frozen shellfish, it is important to verify that the product is frozen and has not been thawed and refrozen. **See Figure 4-40.** If shellfish is thawed on the edges, it should be rejected. Shellfish that has been refrozen will have a poor texture. Dry spots on the shellfish indicate that it has thawed slightly and been refrozen. Ice crystals are another sign that shellfish has been thawed or not stored at appropriate temperatures. Shellfish displaying dry spots or ice crystals should be rejected. Maximum shelf life is obtained by storing frozen shellfish at 0°F or below.

Frozen Shellfish

Fortune Fish Company

Figure 4-40. When receiving frozen shellfish, verify the product is frozen and has not been thawed and refrozen.

CHECKPOINT 4-8

1. Describe various market forms of shellfish.

2. Explain how live shellfish are received and stored.

3. Describe the purpose of a shellstock tag.

4. Explain how frozen shellfish are received and stored.

FABRICATING SHELLFISH

Shellfish are often purchased whole and fabricated in-house. Common shellfish fabrication techniques include deveining shrimp, debearding mussels, splitting lobster tails, cleaning soft-shell crabs, shucking oysters, cleaning squid, and shucking clams.

Deveining Shrimp

Shrimp contain a sand vein, or intestinal tract, that must be removed prior to further preparation. The shells are also removed.

Debearding Mussels

Mussels have hairlike threads called beards that need to be removed prior to preparation. If the beards are removed too soon, the mussels will die.

Procedure for Debearding Mussels

1. Brush the mussel under running water to remove excess mud, sand, and debris.

2. Use a pair of needle-nose pliers to grab the beard and gently pull it free from the mussel.

Splitting Lobster Tails

Broiled lobster tail is a preparation that requires the shell of the tail to be split open. The tail meat is pulled up from, but left inside, the shell for service.

Procedure for Deveining Shrimp

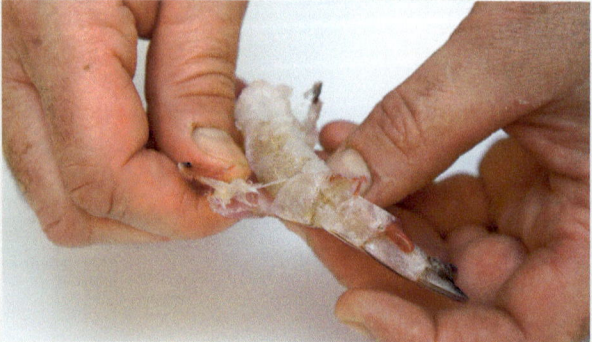 

1. Hold the tail fins in one hand with the underside of the shrimp facing up. Pinch the legs between the thumb and index finger of the other hand and gently pull to remove.

2. Grasp the tail between the thumb and index finger of the guiding hand. Use the other hand to gently grasp the shell and twist it slightly to separate the shell from the body.

3. Use a paring knife to make a shallow slice along the back of the shrimp to expose the sand vein. *Note:* If the shrimp is to be butterflied, make a deeper slice to open the flesh so that it lays flat.

4. Gently pull the sand vein to remove it. Discard the vein. *Note:* Holding the shrimp under cold running water may facilitate the removal of the sand vein.

Deveining Shrimp
Media Clip

Procedure for Splitting Lobster Tails

1. With the underside of the lobster facing down, insert the tip of a chef's knife into the center of the tail just above the bottom fins.

2. Use the guiding hand to push the handle of the knife down to slice through the top portion of the shell lengthwise.

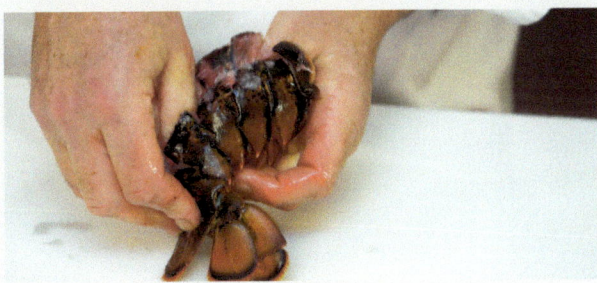

3. Spread the cut shell all the way open to reveal the tail flesh.

4. Gently pull the tail flesh up through the cut shell for presentation.
5. Make a few shallow slices along the underside of the tail flesh.

Cleaning Soft-Shell Crabs

Cleaning soft-shell crabs involves removing the inedible portions, such as the eyes and the apron. **See Figure 4-41.** The apron is tucked under the body and used to carry and conceal eggs.

Soft-Shell Crabs

Eyes

Apron is underneath

Figure 4-41. Cleaning soft-shell crabs involves removing the inedible portions, such as the eyes and the apron.

Nutrition Note

Crabs are a good source of protein, vitamin B12, zinc, and copper. King crabs are also high in vitamin C. However, crabs are very high in cholesterol and sodium.

Procedure for Cleaning Soft-Shell Crabs

1. Peel back the pointed top shell of the crab to reveal the stringlike gills. Use a boning knife to carefully scrape away the gills.
2. Locate the eyes on the top of the head. Use kitchen shears to remove the mouth and head by cutting just behind the eyes.
3. Gently squeeze the body just behind where the head was removed to release a green bubble of fluid. Rinse the green fluid away.
4. Turn the crab over and locate the apron. Firmly twist and then pull the apron to remove it. Discard the apron. *Note:* The intestinal tract will be attached to the apron.

Shucking Oysters

Shucking is the process of opening a bivalve. Bivalves such as oysters and clams must be shucked to access the edible flesh inside the shell. An oyster knife is used to shuck oysters.

Cleaning Squid

Squid is often shipped cleaned and ready for cooking. If a squid is white and the eyes have been removed, it has already been cleaned. If a squid is dark in color and the eyes are intact, it must be cleaned before being prepared. The tentacles may be left whole or cut into smaller sections. The mantle can be cut into rings, sections, or left whole.

Shucking Clams

Clams are shucked with a clam knife. Clams also may be steamed open to access the flesh.

Procedure for Shucking Oysters

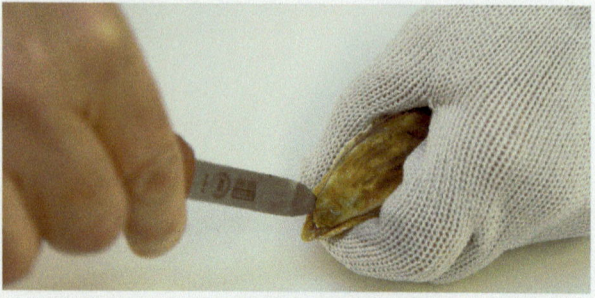

1. While wearing a mesh glove on the guiding hand, firmly hold the oyster in the palm, hinge side up.
2. Insert the tip of an oyster knife near the hinge of the shell.

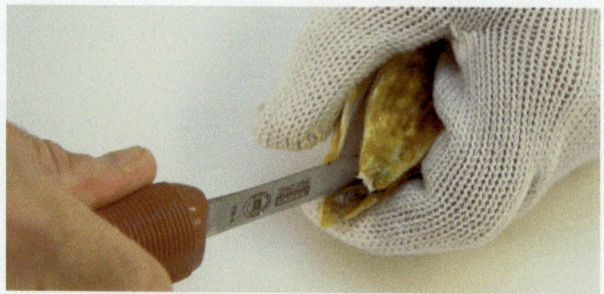

3. Carefully twist the blade to pop the hinge open. *Note:* Care should be taken not to cut the flesh of the oyster.

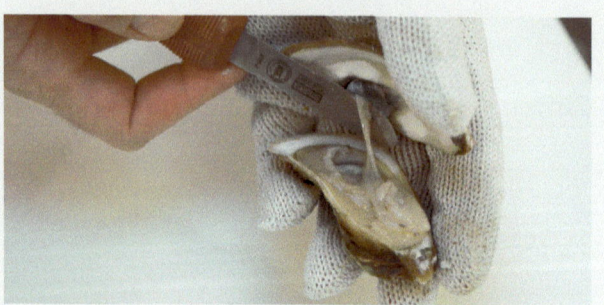

4. Slide the oyster knife under the flesh of the top shell to cut it away from the shell. *Note:* Keep the oyster as level as possible to avoid spilling the juice.

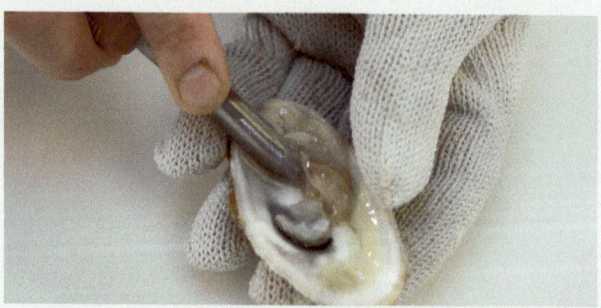

5. Slide the oyster knife under the adductor muscle to separate the flesh from the bottom shell.
6. Remove any shell fragments that may have mixed with the oyster flesh.

Procedure for Cleaning Squid

1. Gently pull the arms and tentacles of the squid away from the body (mantle).
2. Using a boning knife, cut just below each of the eyes to remove the eight arms and two tentacles from the head.
3. Remove the ink sac from the head and reserve.
4. Pull back the tentacles to expose the beak. Carefully pull out the beak and discard.
5. Hold the body (mantle) in guiding hand and carefully pull out the transparent quill with the other hand. Discard the quill.
6. Carefully pull as much skin as possible from the body (mantle) to reveal the lighter-colored flesh below. Discard the skin.
7. Thoroughly rinse all of the squid flesh before use.

Shucking Oysters
Media Clip

Fabricating Squid

Procedure for Shucking Clams

 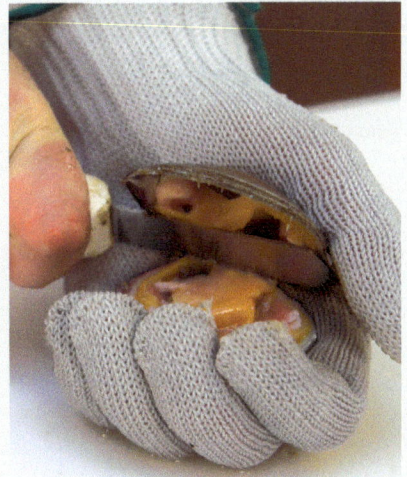 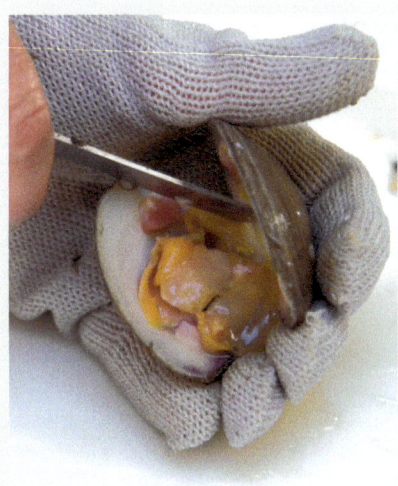

1. While wearing a mesh glove on the guiding hand, firmly hold the clam with the hinge resting against the base of the thumb and the lip facing the fingertips.
2. Insert the blade of a clam knife between the top shell and bottom shell until halfway inserted.
3. Carefully twist the blade to pop the shells open.

4. Use the tip of the clam knife to separate the flesh from the top shell. Discard the top shell.

5. Use the knife blade to separate the adductor muscle from the bottom shell.
6. Remove any shell fragments that may have mixed with the clam flesh.

CHECKPOINT 4-9

1. Devein shrimp.
2. Debeard mussels.
3. Explain how to split lobster tails.
4. Split a lobster tail.
5. Explain how to clean soft-shell crabs.
6. Explain how to shuck oysters.
7. Shuck oysters.
8. Explain how to clean squid.
9. Explain how to shuck clams.
10. Shuck clams.

MARKET FORMS OF BEEF

Beef is the flesh of domesticated cattle. The age and gender of the cattle affects the quality and flavor of the meat.

Most beef comes from steers. A *steer* is a male calf that has been castrated prior to reaching sexual maturity. Steers that are 15–24 months old produce the best-quality beef. Most steers are grain-fed, which increases the marbling of the meat. *Grain-fed beef* is meat from cattle that were grain-fed in confined feeding operations for 90 days to 1 year. *Grass-fed beef* is meat from cattle that were raised on grass or hay with little or no special feed. Grass-fed beef contains less saturated fat than grain-fed beef.

Organic beef is produced within the strict guidelines of and audited by the USDA. There are many restrictions to the production and processing of meat from cattle that is certified as organic.

Market Forms of Beef

In order for the term "natural" to appear on a food label, the USDA requires that the product be minimally processed, not contain any artificial ingredients, and not contain any preservatives. The USDA has no specific restriction on management practices during the life of the animal.

Beef is available in a variety of market forms. Knowledge of each of these market forms is necessary for accurate product ordering. Common market forms of beef include partial carcasses, primal cuts, and fabricated cuts.

fabrication and storage must be available. In addition, foodservice staff capable of cutting a side of beef is required. Because of the amount of space, time, and labor required to break down a side of beef, sides are rarely purchased by foodservice operations.

A *quarter of beef* is a side of beef that has been divided into two parts between the 12th and 13th ribs. Quarters are sold as forequarters and hindquarters. Quarters are not often purchased by foodservice operations.

Partial Carcasses of Beef

Sides are the most common partial carcass of beef. A *side of beef* is a half of a carcass split along the backbone. There are two sides to each carcass. A side can be purchased for less per pound than any primal or fabricated cut. However, in order for purchasing sides of beef to be cost-effective, adequate facilities and equipment for

Primal and Fabricated Cuts of Beef

A *primal cut* is a large cut from a whole or a partial carcass. A single beef carcass has two of each primal cut (a left side and a right side). Primal cuts of beef are very large. The eight primal cuts of beef are the chuck, rib, short loin, sirloin, round, flank, short plate, and brisket and shank. **See Figure 4-42.**

Primal Cuts of Beef

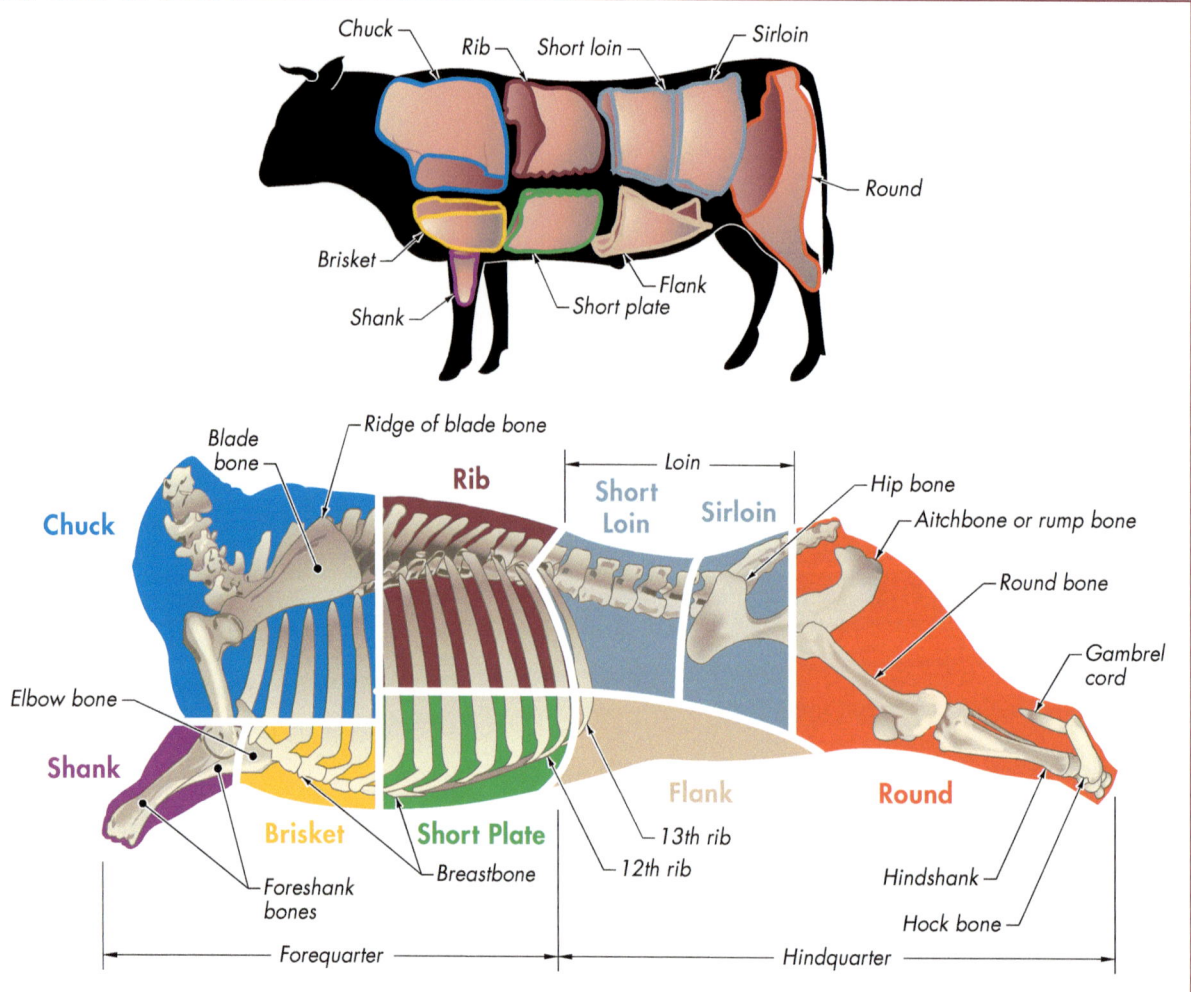

Figure 4-42. The primal cuts of beef include the chuck, rib, short loin, sirloin, round, flank, short plate, and brisket and shank.

Each primal cut of beef is further divided into fabricated cuts. A *fabricated cut* is a ready-to-cook cut that is packaged to certain size and weight specifications. Fabricated beef cuts include short ribs and tenderloins, as well as top sirloin, flat-iron, eye-roll, T-bone, porterhouse, Delmonico, butt, and skirt steaks. Fabricated cuts are a convenient way of providing uniform portions while reducing labor costs. The price per pound for fabricated cuts is higher than the price per pound for primal cuts. Foodservice operations often purchase some primal cuts and some fabricated cuts of beef.

Chucks. A *beef chuck* is a shoulder primal cut of beef that contains the first five rib bones, some of the backbone, and a small amount of the arm and blade bones. The chuck is the largest primal cut and its average weight is approximately 26% of the total carcass weight. The shoulder is one of the most-exercised muscles on the animal, so it is a tough cut of meat with a lot of connective tissue. However, chuck is also quite lean and has an excellent flavor.

Fabricated cuts from the chuck include shoulder clods, clod tenders, chuck rolls, top blade chucks, flat-iron steaks, and strips. **See Figure 4-43.** Flat-iron steaks are the second-most-tender chuck cuts after the tenderloin. Short ribs are produced from the small rib bone ends that are sawed off of the primal rib as the rib roast is removed. Short ribs have a sizable portion of lean meat on them.

Larger pieces of chuck lend themselves to braising and stewing. Smaller pieces of chuck and trimmings produce very flavorful ground meat.

Ribs. A *beef rib* is the primal cut of beef located between the chuck and short loin and contains seven rib bones, from the 6th to the 12th rib. Its average weight is approximately 10% of the total carcass weight. The meat is tender and well marbled. The rib bones (finger bones) make the rib cut a good cut for roasting, because they form a natural rack on which the meat can cook.

A beef rib is often fabricated into a variety of cuts. **See Figure 4-44.** A beef rib contains the prime rib roast. The rib bones can be left on for roasting to produce a moist roast. If the bones are removed, a boneless rib eye roast can be further cut into fabricated bone-in or boneless rib eye steaks. A *rib eye* is a large, eye-shaped muscle within the rib that is a continuation of the sirloin muscle. Meat from eye-shaped muscles, such as the rib eye or tenderloin, is often referred to as eye meat. The 6th to the 12th ribs can be prepared as smoked or barbequed beef ribs. If the rib bones are removed, the meat can be rolled and tied into a rolled-rib roast.

Chuck Cuts of Beef

Shoulder Clod

Clod Tender

Chuck Roll

Top Blade Chuck

Flat-Iron Steaks

Strips

Canadian Beef, Beef Information Centre

Figure 4-43. A beef chuck is often fabricated into a variety of cuts including shoulder clods, clod tenders, chuck rolls, top blade chucks, flat-iron steaks, and strips.

Rib Cuts of Beef

Rib Roast

Back Ribs

Prime Rib

Trimmed Rib Roast

Bone-In Rib Eye Steaks

Boneless Rib Eye Steaks

Canadian Beef, Beef Information Centre

Figure 4-44. A beef rib is often fabricated into a variety of cuts.

Short Loins. A *beef short loin* is a primal cut of beef located just to the rear of the primal rib and includes the 13th rib and a small section of the backbone. Its average weight is approximately 8% of the total carcass weight. The short loin can be cut in cross sections to produce some of the most popular fabricated cuts. **See Figure 4-45.** Cuts from the short loin are commonly grilled, broiled, or roasted.

When the short loin and sirloin are split apart, the smaller portion of tenderloin is part of the short loin. A *beef tenderloin* is an eye-shaped muscle running from the primal rib cut into the primal leg. The tenderloin is located just beneath the strip loin and is the most tender piece of beef. Sometimes the entire tenderloin is removed prior to dividing the short loin and sirloin. The whole tenderloin can be roasted whole or divided into Chateaubriand, filets mignons, and tournedos.

A *beef strip loin* is a short loin without a tenderloin. A strip loin can be cut into boneless strip steaks or roasted whole. Fabricated cuts from the short loin are often aged, as they have ample fat covering and marbling and are very tender. Aging beef loin intensifies the flavor and tenderness of the meat.

The short loin produces many popular fabricated cuts. Starting from the end nearest the primal rib, cross-section cuts produce Delmonico steaks, T-bone steaks, and porterhouse steaks. Delmonico steaks do not include any tenderloin, T-bone steaks include only a small section of tenderloin, and porterhouse steaks include a large section of tenderloin.

Sirloins. A *beef sirloin* is a primal cut of beef situated just behind the short loin and contains some of the backbone and hip bone. Its average weight is approximately 9% of the total carcass weight. With the exception of the butt tenderloin muscle, meat from the sirloin is not quite as tender as meat from the short loin. However, the sirloin can be cut into butt steaks that can be marinated, skewered, and then grilled or broiled. Ball-tip steaks, sirloin flaps, sirloin tri-tips, sirloin tri-tip steaks, top sirloin roasts, top sirloin steaks, top sirloin caps, and top sirloin cap steaks can be fabricated from the sirloin primal cut. **See Figure 4-46.**

Rounds. A *beef round* is a primal cut of beef that includes a large grouping of muscles that represent the hind hip and thigh of the carcass. The average weight of the round is approximately 27% of the total carcass weight. A round contains large bones including the round bone (leg), aitchbone (pelvis), shank, and tailbone.

Short Loin Cuts of Beef

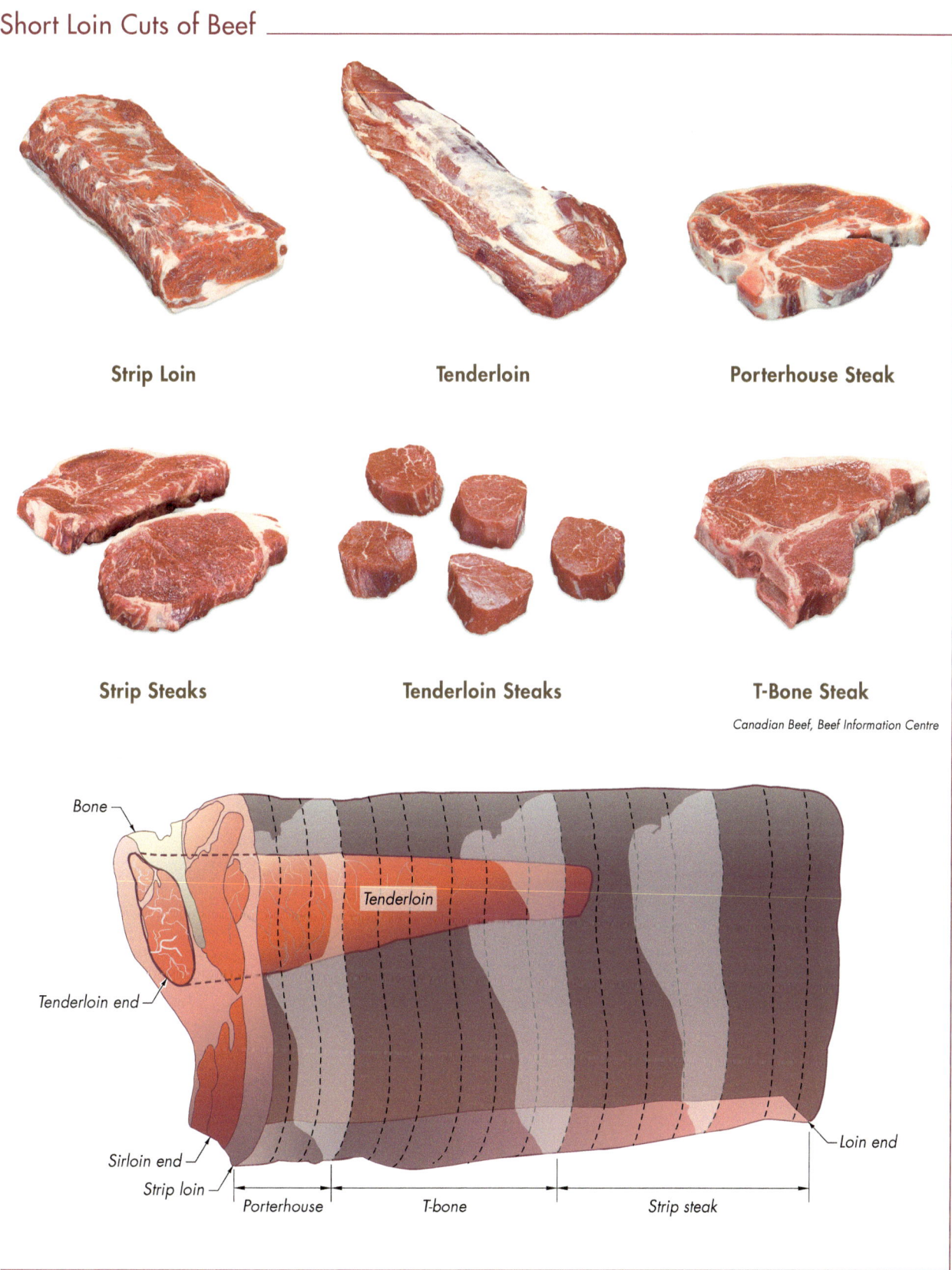

Strip Loin

Tenderloin

Porterhouse Steak

Strip Steaks

Tenderloin Steaks

T-Bone Steak

Canadian Beef, Beef Information Centre

Figure 4-45. The short loin can be cut in cross sections to produce some of the most popular fabricated cuts.

Sirloin Cuts of Beef

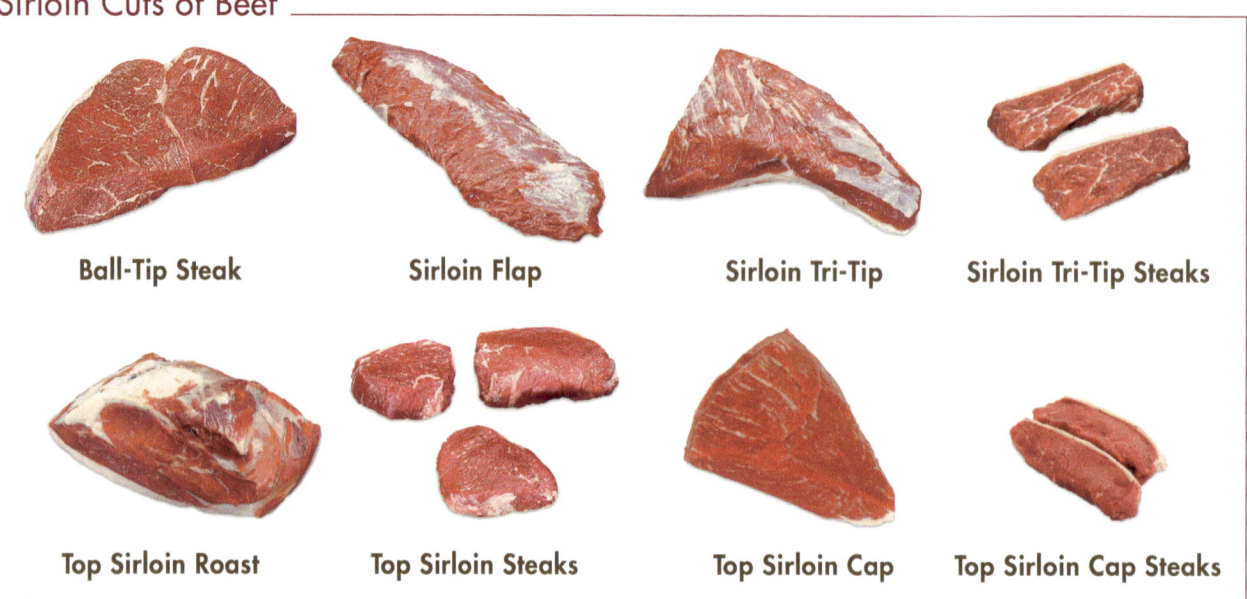

Canadian Beef, Beef Information Centre

Figure 4-46. Ball-tip steaks, sirloin flaps, sirloin tri-tips, sirloin tri-tip steaks, top sirloin roasts, top sirloin steaks, top sirloin caps, and top sirloin cap steaks can be fabricated from the sirloin primal cut.

A round can be slow-roasted whole. However, due to its size, the round is commonly broken down and sold as separate subprimal cuts. Subprimal and fabricated cuts from the round include the top round, bottom round, knuckle, and shank. The bottom round can be cut into the outside round and the eye of round. **See Figure 4-47.**

Round Cuts of Beef

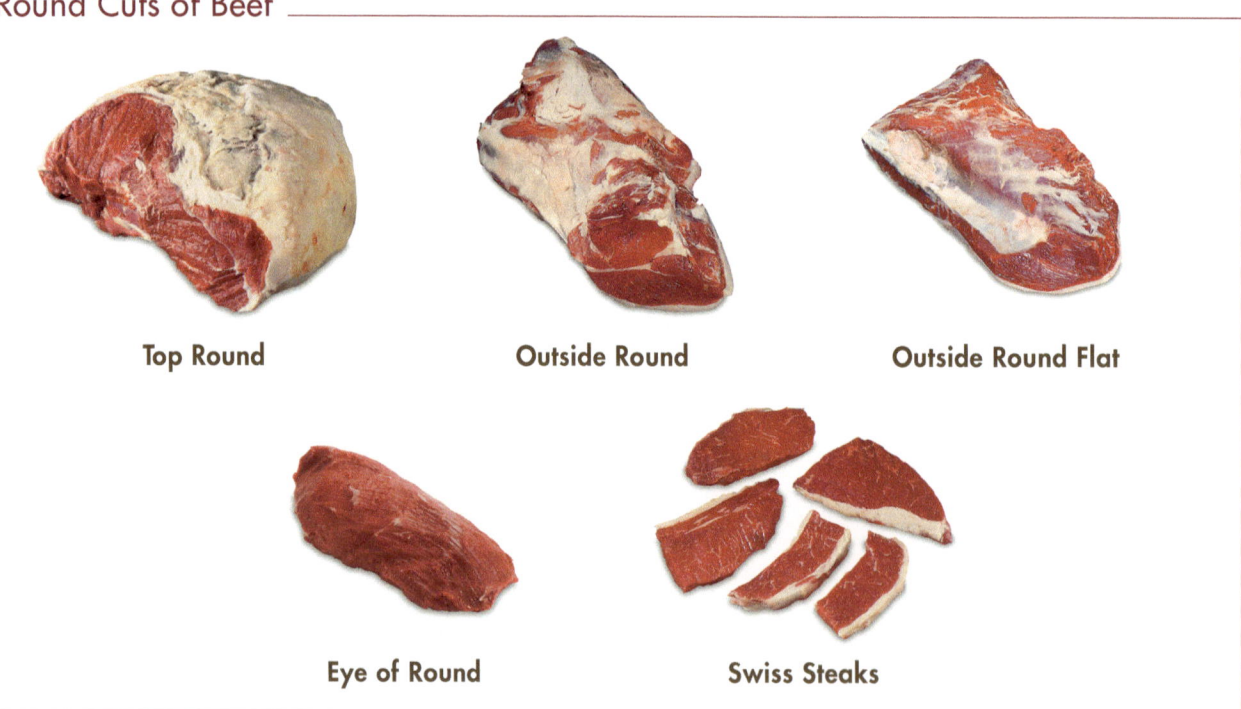

Canadian Beef, Beef Information Centre

Figure 4-47. The round can be broken down into a variety of subprimal and fabricated cuts.

When the round is trimmed for cooking it yields a large amount of usable meat. A *beef rump roast* is a roast cut from the primal round, above the back end of the hip bone. If the bone is left in, it is called a standing rump roast. A boneless rump roast is rolled and tied prior to being sold. The rump roast is a very flavorful cut that is often braised. A *steamship round roast* is the beef round with the shank and rump removed. The top round and knuckle portions can be roasted. The outside round and eye of round are often braised.

Flanks. The *beef flank* is a primal cut of beef that includes the thin, flat section of the hindquarters located beneath the loin. Its average weight is approximately 4% of the total carcass weight. The flank has more fat than lean meat and contains one thin, oval-shaped, boneless flank steak. The flank steak can be scored or cubed before it is cooked. Prior to cooking, the fat covering should be removed and the meat can be marinated to produce a more tender and flavorful piece of meat. If cooked whole, flank steak should be cut across the grain or it will be tough. **See Figure 4-48.**

Flank Cuts of Beef

Canadian Beef, Beef Information Centre

Figure 4-48. A beef flank is fabricated into flank steaks.

Short Plates. A *beef short plate* is a primal cut of beef that includes a thin portion of the beef forequarter located just beneath the rib cut. The average weight of the short plate is approximately 5% of the total carcass weight. The bones attached to this cut are the remaining sections of the rib bones. The small bones from the short plate are called short ribs. Short ribs are not quite as meaty as those from the rib cut. In addition to short ribs, the short plate yields the flavorful skirt steak. **See Figure 4-49.**

Short Plate Cuts of Beef

Outside Skirt Steak

Inside Skirt Steak

Short Ribs

Canadian Beef, Beef Information Centre

Figure 4-49. A beef short plate is fabricated into skirt steak and short ribs.

Briskets and Shanks. The brisket and shank are two separate muscle groups that make up one primal cut of beef that is located just below the chuck. **See Figure 4-50.** The weight of the brisket and shank is approximately 9% of the total carcass weight. The *beef brisket* is a thin section of beef that contains some of the ribs, the breastbone, and layers of lean muscle, fat, and connective tissue. The ribs and breastbone are always removed prior to cooking the brisket.

Brisket and Shank Cuts of Beef

Brisket

Ground Shank   **Stew Beef**

Canadian Beef, Beef Information Centre

Figure 4-50. A brisket and shank are two separate muscle groups that make up one primal cut of beef that is located just below the chuck.

Brisket is a tough cut of beef with excellent flavor that can become tender when cooked properly. It has long muscle fibers that run in several directions, making it difficult to slice. Brisket is often braised or simmered. Brisket can be cured, peppered to make pastrami, braised as sauerbraten, simmered as New England-style brisket, and corned or pickled as corned beef.

The *beef shank* is a bony section of beef that is surrounded by a small amount of very tough but flavorful meat. Shanks are used for making stocks and rich reduction sauces. Shank meat is usually ground to flavor and clarify consommés, because the meat has a high concentration of collagen that converts to gelatin when cooked. Shanks are generally cut perpendicular across the bone and then braised.

Beef Offals

An *offal* is an edible part of an animal that is not part of a primal cut. Commonly used beef offals include liver, tongue, tripe, oxtail, and kidney. **See Figure 4-51.** With the exception of liver, beef offals are prepared using moist-heat cooking methods. Beef liver is covered with a thin membrane that should be removed before slicing. Beef liver is commonly broiled, sautéed, or pan-fried.

Beef Offals

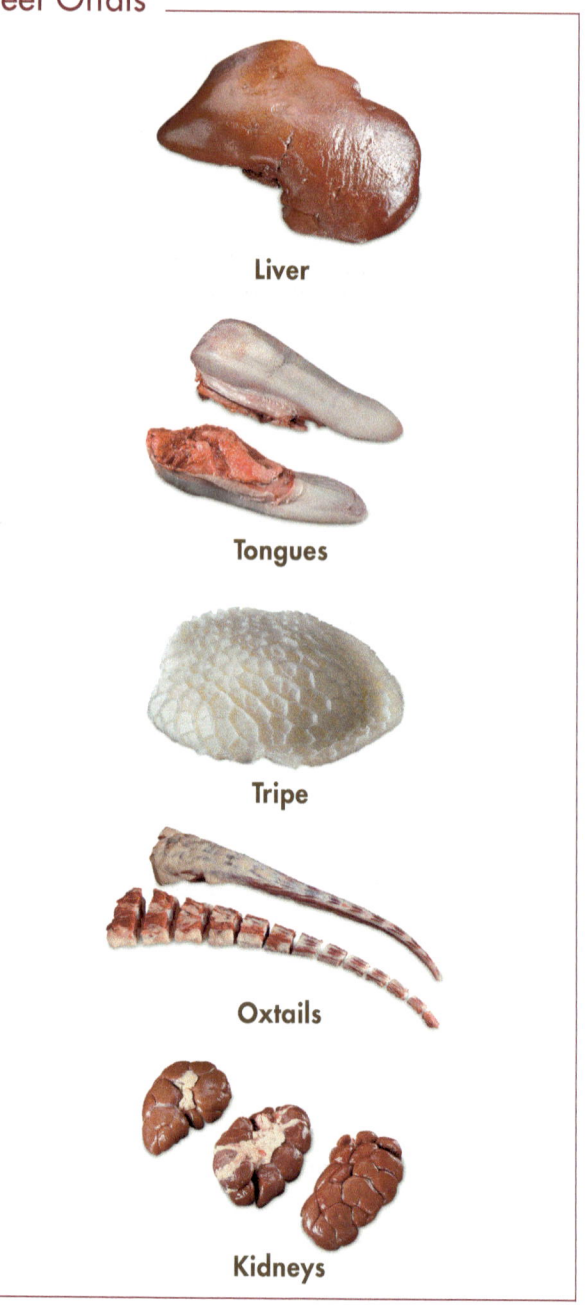

Liver

Tongues

Tripe

Oxtails

Kidneys

Canadian Beef, Beef Information Centre

Figure 4-51. Beef offals include liver, tongue, tripe, oxtail, and kidney.

Beef tongue is available fresh, smoked, pickled, and corned, but smoked is the most popular form. Tongue is always prepared using moist-heat cooking methods. To determine doneness, the tip of the tongue is felt to test tenderness. When the tip is soft, the tongue is done. Cooked tongue is cooled in cold water and then skinned. Beef tongue can be served cold or can be reheated and served hot. **See Figure 4-52.**

Beef Tongues

Figure 4-52. Beef tongue can be served cold or can be reheated and served hot.

Tripe is the muscular inner lining of a stomach of an animal, such as cattle or sheep. *Honeycomb tripe* is the lining of the second stomach found in cattle. Honeycomb tripe may be purchased fresh, pickled, or canned. It is often fried, creamed, used in soups, or served cold with a vinaigrette dressing.

An *oxtail* is the tail from a cattle carcass. Oxtail contains a considerable amount of bone and a good portion of richly flavored flesh. Oxtail is most often used in stews or braised in rich cooking liquids. The thin end of the oxtail can be used to make oxtail soup.

Beef kidneys have irregularly shaped lobes divided by deep cracks. Before cooking, the kidney is split lengthwise, and the suet (fat) and urinary canals are carefully removed. Kidneys are commonly braised.

CHECKPOINT 4-10

1. Identify the eight primal cuts of beef.

2. Describe the advantages and disadvantages of purchasing fabricated cuts.

3. Identify the cuts fabricated from each primal cut of beef.

4. Explain how to prepare beef offals.

Photo Courtesy of D'Artagnan, Photography by Doug Adams Studio

MARKET FORMS OF VEAL

Veal is the flesh of calves, which are young cattle. Veal is available in a variety of market forms, including whole and partial carcasses, primal cuts, fabricated cuts, and offals. Knowledge of these different market forms is necessary for accurate product ordering. Understanding the skeletal structure of calves can aid in identifying the market forms of veal. **See Figure 4-53.**

Nutrition Note

Veal is a good source of protein and is low in fat. However, veal contains more cholesterol than beef.

Partial Carcasses of Veal

In order for the purchasing of partial carcasses of veal to be cost-effective, skilled labor and storage space are required. Veal is typically not split into sides like beef. Veal is split into head and tail sections known as the foresaddle and hindsaddle. The foresaddle and hindsaddle are split between the 11th and 12th ribs, not down the backbone. A *veal foresaddle* is the front half of a carcass consisting of the primal shoulder, rack, breast, and shank cuts. A *veal hindsaddle* is the rear half of a carcass consisting of the loin and leg.

Primal and Fabricated Cuts of Veal

The left and right primal cuts of veal remain joined together and are sold as a single cut. For example, the leg primal cut has two joined legs. The primal cuts of veal include the shoulder, rack, loin, leg, and foreshank and breast. **See Figure 4-54.**

Veal Facts

Skeletal Structure of Calves

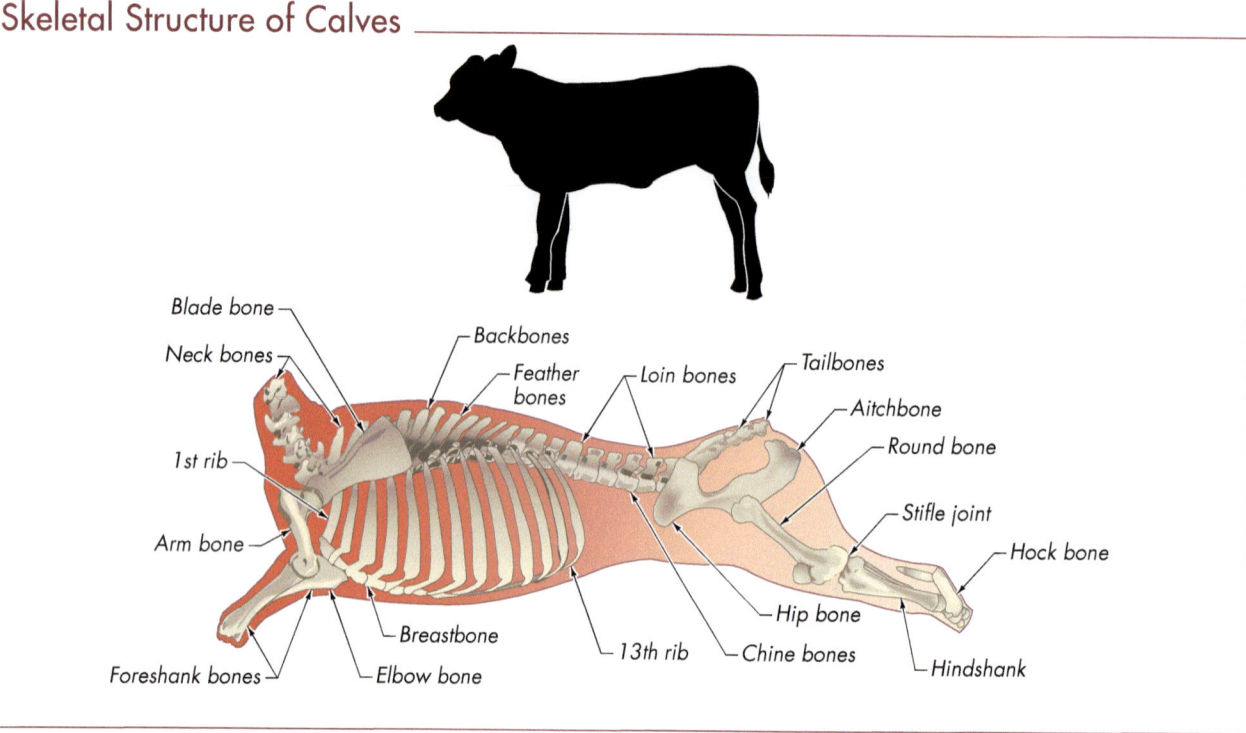

Figure 4-53. Understanding the skeletal structure of calves can aid in identifying the market forms of veal.

Primal Cuts of Veal

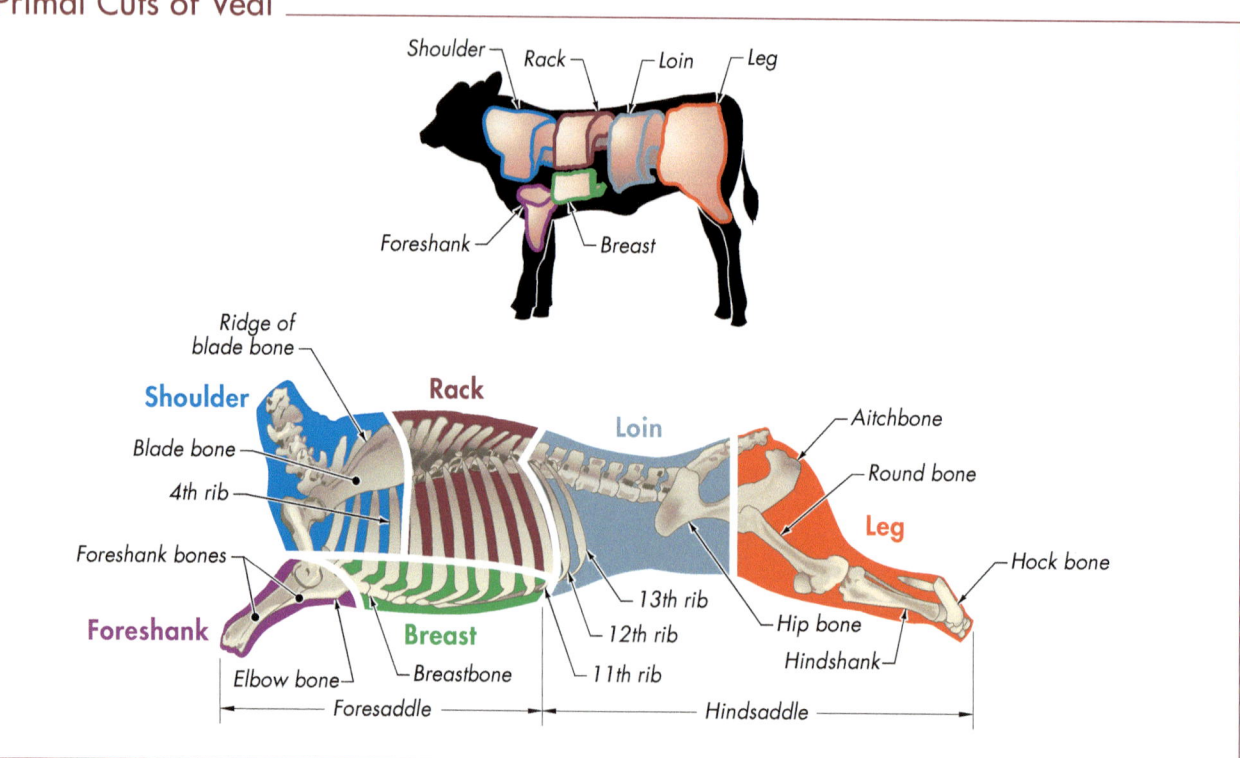

Figure 4-54. The primal cuts of veal include the shoulder, rack, loin, leg, and foreshank and breast.

Each primal cut is further divided into fabricated cuts. Fabricated cuts of veal include frenched chops, baby T-bone steaks, boneless cutlets, and ossobuco-cut shanks. Some foodservice operations may choose to purchase only fabricated cuts of veal. However, the price per pound is higher for fabricated cuts than the price per pound for primal cuts.

Shoulders. A *veal shoulder* is a primal cut that contains the first four rib bones, some of the backbone, and a small amount of the arm and blade bones. The average weight of a veal shoulder is approximately 21% of the total carcass weight. Veal shoulder is tough but flavorful. It can be fabricated into steaks or chops. However, it is most often ground, cut into cubes for stewing, or cooked whole. **See Figure 4-55.**

Racks. The *veal rack* is a primal cut located between the shoulder and loin and contains seven rib bones. Its average weight is approximately 9% of the total carcass weight. The meat is tender and well marbled. A veal rib is different from a beef rib in that veal is not split into two halves along the backbone (2nd chine bones). An unseparated veal rack is called a hotel rack and consists of two very tender veal rib loins. A veal rack can be split into halves and tied into a circle to form a crown rib roast.

Veal racks can be trimmed, frenched, and cut into veal chops. **See Figure 4-56.** Other fabricated cuts from the rack include a small portion of the tenderloin, known as the short tenderloin, and the boneless veal rib eye roast.

Loins. A *veal loin* is a primal cut located between the primal rack and leg and includes the 12th and 13th rib, the loin eye muscle, the center section of the tenderloin, the strip loin, and flank meat. The average weight of the loin is approximately 10% of the total carcass weight. A complete, unsplit primal loin from a veal carcass is commonly referred to as a saddle. Veal loins are often divided into fabricated cuts such as baby T-bones and cutlets. **See Figure 4-57.**

Shoulder Cuts of Veal

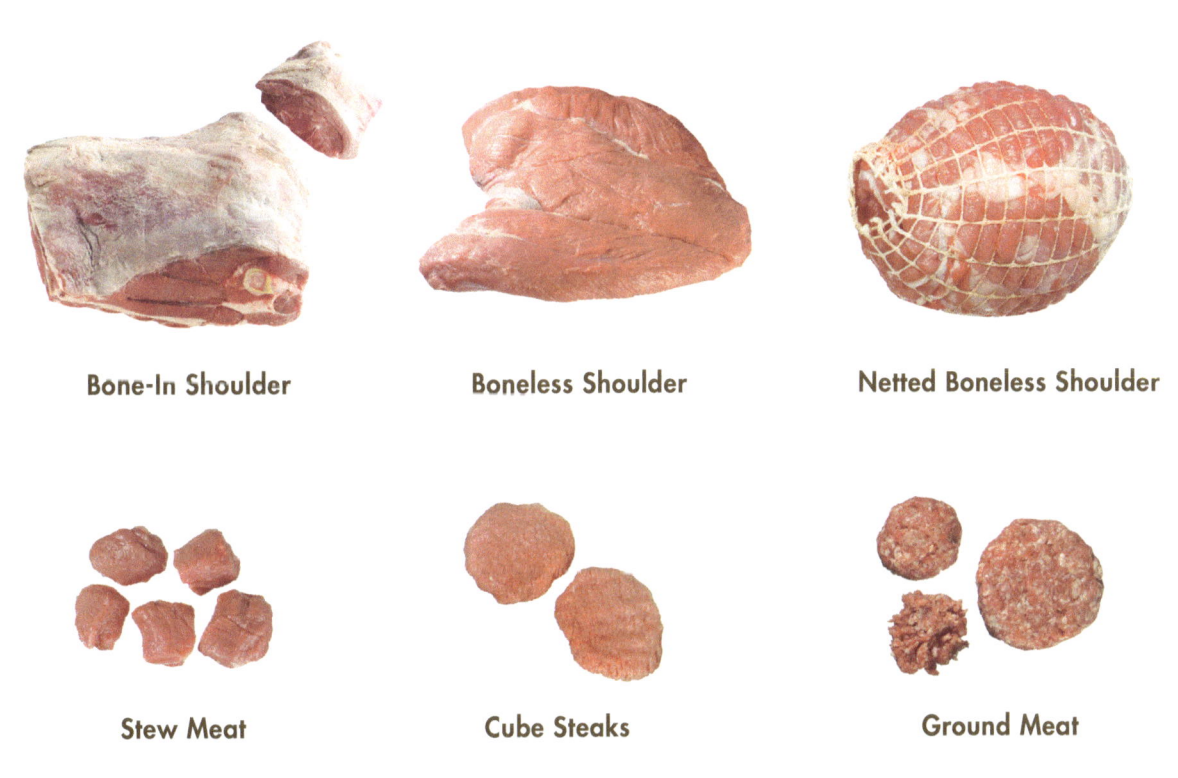

Bone-In Shoulder **Boneless Shoulder** **Netted Boneless Shoulder**

Stew Meat **Cube Steaks** **Ground Meat**

Strauss Free Raised

Figure 4-55. A veal shoulder can be fabricated into steaks or chops. However, it is most often ground, cut into cubes for stewing, or cooked whole.

Rack Cuts of Veal

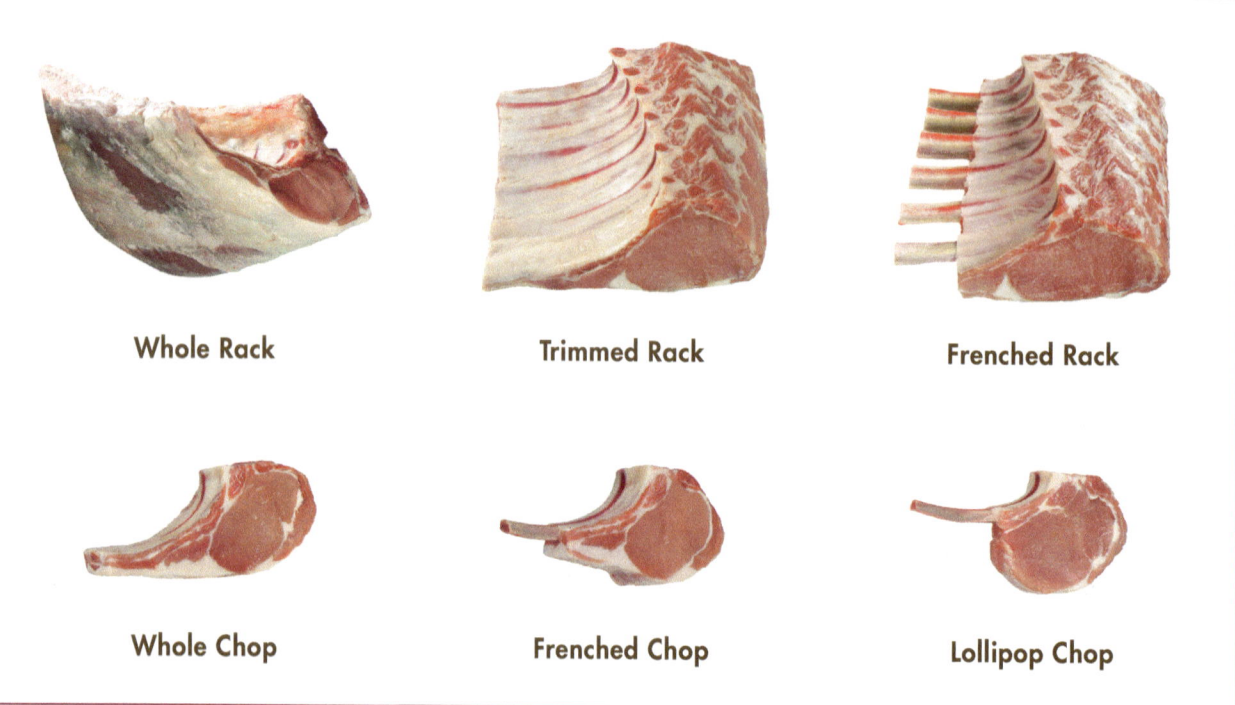

Whole Rack   Trimmed Rack   Frenched Rack

Whole Chop   Frenched Chop   Lollipop Chop

Strauss Free Raised

Figure 4-56. Veal racks can be trimmed, frenched, and cut into veal chops.

Loin Cuts of Veal

Trimmed Short Loin   Strip Loins   Loin Chops

Tenderloin   Filet Mignons   Porterhouse Steaks

Strauss Free Raised

Figure 4-57. Veal loins are sometimes roasted whole, however they are often divided into fabricated cuts.

A *baby T-bone* is a 6–8 oz steak cut from the loin of veal. It contains loin meat on one side and tenderloin on the other side. Cuts from the loin are usually grilled, broiled, or roasted. A *veal cutlet* is a thin slice of veal. Wiener schnitzels are prepared by pounding veal cutlets until very thin and then breading and frying them. Wiener schnitzels are commonly served with lemon, topped with a variety of vegetables, or topped with woodland cream sauce.

Legs. A *veal leg* is a primal cut from the hind leg that contains the leg, sirloin, last portion of the backbone, pelvis (hip bone and aitchbone), round bone, hindshank, and tailbone. Its average weight is approximately 42% of the total carcass weight. The leg is the most versatile cut of veal because it contains solid, lean, fine-textured meat. Tender meat is located near the sirloin end, and tougher meat is located toward the shank. The entire leg is typically boned and cut into scallops or cutlets rather than being roasted whole. The leg is boned by following the muscle structure of the meat so that pieces of equal tenderness are removed.

Veal legs can be divided into leg, hindshank, ossobuco, inside round, eye round, and scallopini cuts. **See Figure 4-58.** These cuts are commonly sliced against the grain and pounded until thin to tenderize them. Ossobuco-cut shanks are cut perpendicular to the bone. A *scallopini* is a small, ¼ inch thick slice of veal (generally leg meat) that is 2–3 inches in diameter.

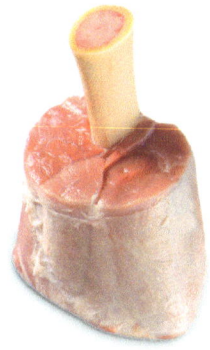

Strauss Free Raised

Foreshanks and Breasts. The foreshank and breast form the primal cut of veal from the lower foresaddle. **See Figure 4-59.** Its average weight is approximately 16% of the total carcass weight. A *veal foreshank* is the upper portion of the front leg of a calf. It can be braised whole or sliced in cross sections across the bone. A *veal breast* is a thin, flat cut of meat located under the shoulder and ribs and contains the breastbone, tips of the rib bones, and cartilage. The breastbone is typically still cartilage because the animal is so young. The thin, flat shape makes the breast easy to stuff, roll, and tie into a tender rolled roast that can be braised to break down the connective tissue.

Leg Cuts of Veal

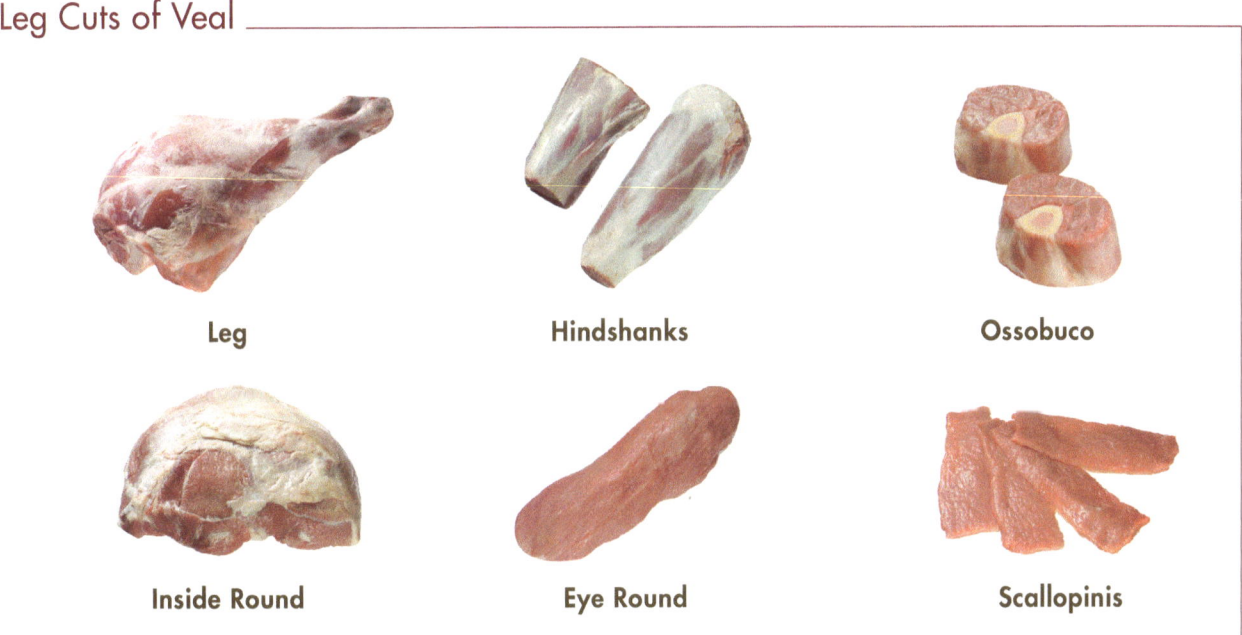

Leg

Hindshanks

Ossobuco

Inside Round

Eye Round

Scallopinis

Strauss Free Raised

Figure 4-58. Veal legs can be divided into leg, hindshank, ossobuco, inside round, eye round, and scallopini cuts.

Foreshank and Breast Cuts of Veal

| Whole Foreshank | Ossobuco-Cut Shanks | Breast |

Strauss Free Raised

Figure 4-59. The foreshank and breast form the primal cut of veal from the lower foresaddle.

Veal Offals

Veal offals are more tender than beef offals. Common veal offals include sweetbreads, liver, and kidney. *Sweetbreads* are the thymus glands of a calf, located in the neck. High-quality sweetbreads should be plump and somewhat firm with a thin protective membrane. Sweetbreads are often sautéed in brown butter or pan-fried. **See Figure 4-60.**

Veal Offals

Sweetbreads

Liver

Kidney

Strauss Free Raised

Figure 4-60. Common veal offals include sweetbreads, liver, and kidney.

Veal liver is commonly broiled, pan-fried, or sautéed and should be cooked medium. Veal kidneys can be broiled with excellent results, unlike beef kidneys, which must be cooked with moist heat. Veal kidneys are typically used to prepare entrées such as kidney stew and kidney pie.

CHECKPOINT 4-11

1. Identify the five primal cuts of veal.

2. Identify the cuts fabricated from each primal cut of veal.

3. Explain how to prepare veal offals.

RECEIVING AND STORING BEEF AND VEAL

Beef and veal are potentially hazardous foods that must be checked for color, odor, texture, and temperature upon receipt. **See Figure 4-61.** Beef should be red in color with white fat. Veal should be pink in color with white fat. There should be no odor, and the texture should be firm and not dry or slick. Refrigerated beef and veal should maintain an internal temperature of 41°F or below. If the meat received is vacuum-packed, its temperature can be taken by placing an instant read thermometer between two of the packages. Beef or veal that is in the temperature danger zone should be rejected.

Receiving Beef and Veal

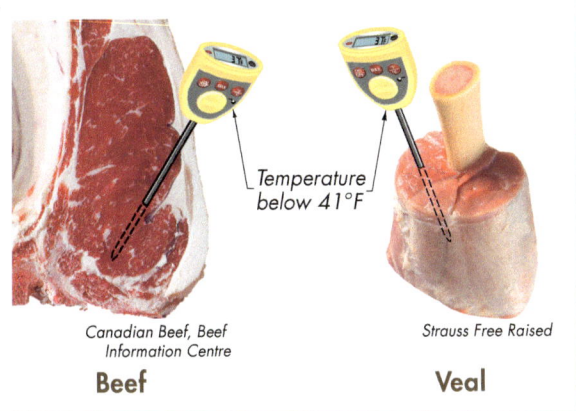

Temperature below 41°F

Canadian Beef, Beef
Information Centre

Beef

Strauss Free Raised

Veal

Figure 4-61. Beef and veal are potentially hazardous foods that must be checked for color, odor, texture, and temperature upon receipt.

Frozen beef and veal should be kept at temperatures that will allow the meat to remain frozen. When frozen meats need to be thawed, they should be placed in the refrigerator overnight. Larger cuts of meat may take more than one day to thaw under refrigeration.

Vacuum-packed beef and veal should never be opened until needed for service or preparation and can be stored refrigerated for three to four weeks. **See Figure 4-62.** Once the vacuum seal is broken, meat has a shelf life of only two to three days. Cut meats should be rewrapped airtight, refrigerated immediately, and used as soon as possible.

Vacuum-Packed Beef

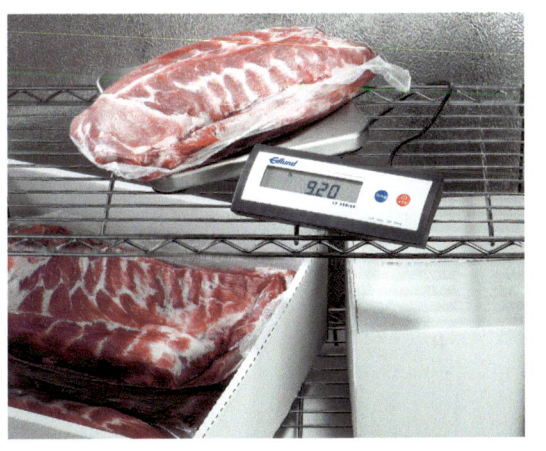

Edlund Co.

Figure 4-62. Vacuum-packed beef and veal should never be opened until needed for service or preparation and can be stored refrigerated for three to four weeks.

All beef and veal used in foodservice operations must be procured from USDA-inspected plants. At the time of slaughter, the carcasses are stamped to indicate that the animal was slaughtered at a USDA-inspected plant. The USDA inspection stamp will be directly on large cuts or on the packaging of fabricated cuts of beef and veal. **See Figure 4-63.** The number on the stamp identifies the plant where the animal was processed. It does not indicate quality.

USDA Inspection Stamps

United States Department of Agriculture

**Stamp for
Whole Carcass**

**Stamp for
Fabricated or
Processed Meats**

Figure 4-63. The USDA inspection stamp will be directly on large cuts or on the packaging of fabricated cuts of beef and veal.

Unlike inspection, USDA quality and yield grading is optional for beef and veal producers. Beef can be graded for quality, yield, or both. Veal can only be graded for quality. Quality and yield grading stamps are stamped on carcasses in the same manner as inspection stamps.

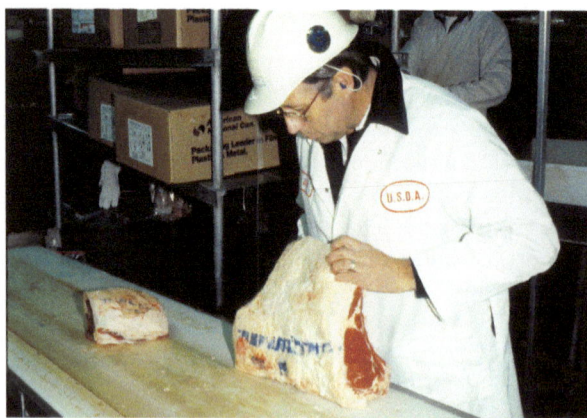

United States Department of Agriculture

Quality Grading

USDA quality grading of beef and veal is voluntary. Beef that has been inspected but not quality graded is referred to as "no roll" beef. Although the grading for beef and veal is based on specific details, the overall terms have relatively the same criteria. However, the quality grade is not a guarantee of quality.

Most veal is USDA Prime or USDA Choice. The grades of beef most commonly used in foodservice operations are USDA Prime, USDA Choice, and USDA Select. **See Figure 4-64.**

USDA Quality Grade Stamps

Figure 4-64. USDA Prime, USDA Choice, and USDA Select beef is commonly used in foodservice operations.

USDA Prime. USDA Prime beef is well marbled and has a thick, firm fat covering. It is the juiciest and most flavorful of all meats, but it is also the most expensive because of the large amount of fat that must be trimmed from it before cooking. Of all the beef marketed in the United States, only a small percentage is graded USDA Prime.

USDA Choice. USDA Choice has slightly less marbling and fat covering than USDA Prime. Most foodservice operations prefer this grade because there is less waste.

USDA Select. USDA Select beef has minimal marbling and a soft fat covering. This grade of beef is often used in quick service operations.

Yield Grading

Yield grades of beef are numbered 1 to 5 and indicate how much usable meat can be obtained from a carcass. A grade of 1 indicates the highest yield of meat, and a grade of 5 indicates the lowest yield of meat. **See Figure 4-65.**

USDA Yield Grades

Grading Beef

Figure 4-65. Most high-quality beef will have a yield grade of 3 or higher.

CHECKPOINT 4-12

1. Identify four traits to check upon receiving beef and veal.

2. Identify the temperature at which refrigerated beef and veal must be kept.

3. Explain why vacuum-sealed packages should only be opened at the time of use.

4. Explain the significance of the USDA inspection stamp.

5. Explain why the quality grade of veal is less important than the quality grade of beef.

6. Describe common USDA quality grades of beef.

FABRICATING BEEF AND VEAL

The size of storage facilities and availability of staff with the fabrication skills often determine whether a foodservice operation purchases primal cuts, fabricated cuts, or a combination. Beef and veal fabrication require trimming, cutting, tenderizing, and tying.

Trimming and Cutting Beef Tenderloin

The tenderloin is often trimmed and then cut into portion-controlled cuts. These cuts include tenderloin tips, Chateaubriand, filets mignons, and tournedos.

Procedure for Trimming and Cutting Beef Tenderloin

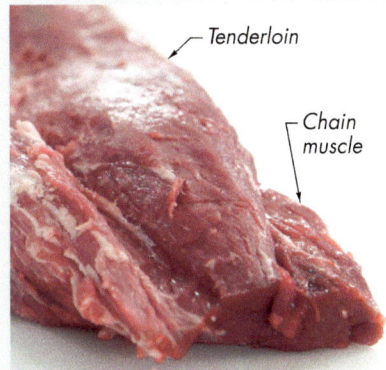
Tenderloin
Chain muscle

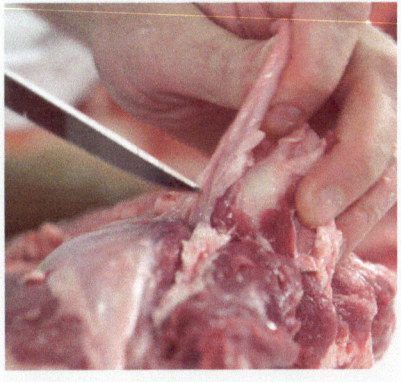

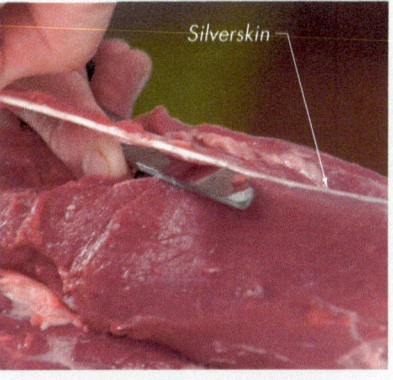
Silverskin

1. With a rigid boning knife, carefully remove the chain muscle from the side of the tenderloin and reserve.

2. Trim and pull the thick fat covering away from the tenderloin.

3. Insert the tip of the boning knife just beneath the silverskin at the tail end of the tenderloin. Draw the blade slightly upward along the length of the tenderloin, just beneath the silverskin, toward the head of the tenderloin.

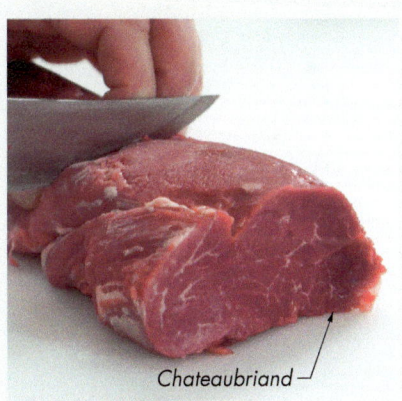
Chateaubriand

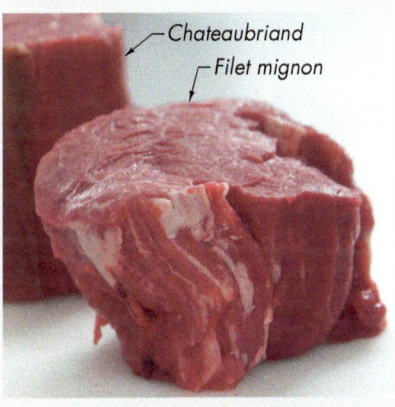
Chateaubriand
Filet mignon

4. Starting at the largest end, cut off the uneven tip of the tenderloin. Cut the tip across the grain into tenderloin tips.

5. Make a cut across the grain just after the large portion ends to remove the Chateaubriand.

6. Cut the center of the tenderloin across the grain to desired thickness to produce filet mignons.

7. Cut the smallest third of the tenderloin across the grain to produce tournedos ½–¾ inch thick and approximately 2½ inches in diameter.

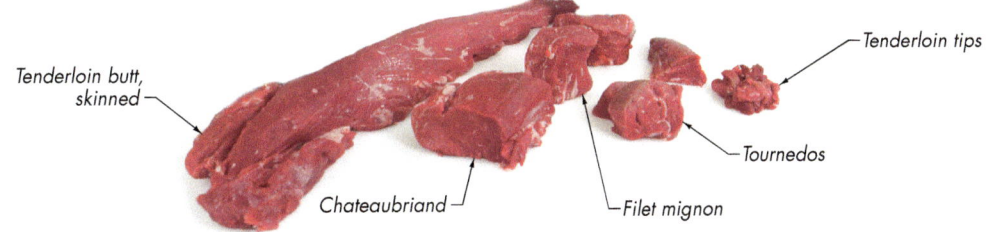
Tenderloin butt, skinned
Chateaubriand
Filet mignon
Tournedos
Tenderloin tips

Fabricated beef tenderloin cuts.

Trimming Beef Tenderloin
Media Clip

Cutting Boneless Strip Loin into Steaks

A boneless strip loin is often trimmed and cut into portion-controlled cuts. The sirloin end of the strip loin contains connective tissue that does not break down during cooking. Steaks cut from this end of the loin are referred to as vein steaks. Although the meat is as tender as the rest of the loin, the amount of connective tissue makes vein steaks less desirable than steaks from the rest of the strip loin.

Procedure for Cutting Boneless Strip Loin into Steaks

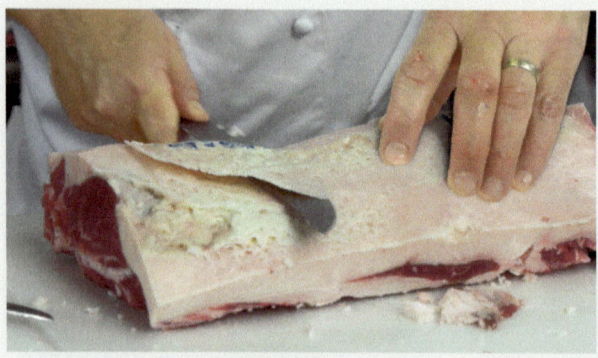

1. Trim the surface fat to approximately ¼ inch thickness.
2. Turn the loin over and trim off any additional fat or connective tissue that could affect the quality of the steaks.

3. Cut the loin across the grain into steaks of desired weight or thickness.

Frenching Veal Chops

Veal chops are a popular menu item and are available boneless or with the bone in. Bone-in veal chops are commonly frenched for service. **See Figure 4-66.**

Frenched Veal Chops

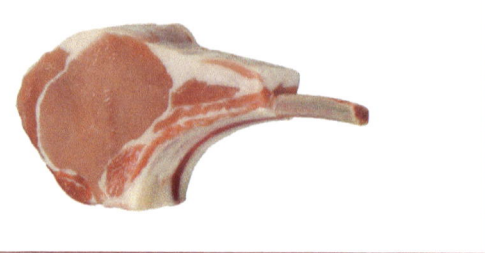

Strauss Free Raised

Figure 4-66. Bone-in veal chops are commonly frenched for service.

Procedure for Frenching Veal Chops

1. Peel back the fat cap where it still is connected to the rack and remove it completely.
2. Trim the entire rack of any excess fat, leaving a thin covering between ¼–⅛ inch thick.
3. Starting on the top side, score the meat by making a straight cut all the way to the bone, approximately 1 inch above the eye meat.
4. Turn the rack over and use the tip of the knife to score the thin membrane on the back of each rib bone from the location of the cut to the tip of the bone.
5. From the scored mark, scrape the meat away from the top of the ribs to expose the bones.
6. Cut between each rib down to the scored mark to remove the meat.

Tenderizing Beef

Beef often benefits from tenderizing. Pounding or cutting meat tenderizes the muscle and connective tissue. **See Figure 4-67.** The following methods are used to tenderize meat:

- pounding meat with a mallet to break up the protein structure and muscle tissue
- using a hand tenderizer with needlelike knives that pierce and gently cut the connective tissues and muscle fibers
- slicing the meat across the grain to produce thin slices with shorter muscle fibers
- grinding the meat to completely break apart strands of connective tissue and muscle tissue

Cutting Boneless Strip Loin
Media Clip

Tenderizing Meat

Pounding with a Mallet

Using a Hand Tenderizer

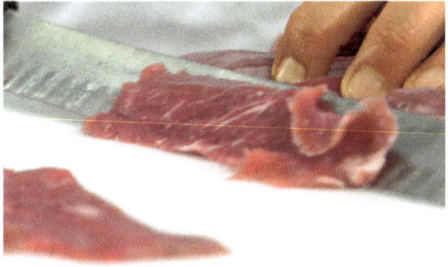

Slicing Across the Grain

Figure 4-67. Pounding or cutting meat tenderizes the muscle and connective tissue.

Ground beef and veal can be used to make meatballs, sausages, meatloafs, and meat fillings. Meat that is to be ground should be very cold before grinding, as should all parts of the grinder that will contact the meat. Room temperature meat and meat that is only somewhat cool has a tendency to be smashed while grinding, which causes it to lose its texture and appear puréed instead of ground. When grinding meat, it should be passed through progressively smaller dies on the grinder in order to evenly distribute the fat and give the mixture a smooth appearance. **See Figure 4-68.**

Grinding Meats

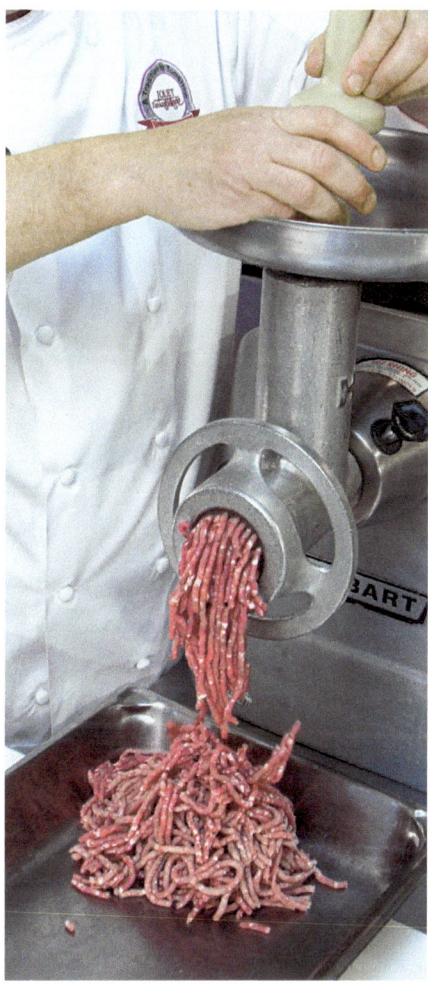

Figure 4-68. When grinding meat, it should be passed through progressively smaller dies on the grinder in order to evenly distribute the fat and give the mixture a smooth appearance.

Tying Beef and Veal

Tying beef and veal helps maintain a consistent shape and ensures that the meat will cook evenly without falling apart. Some cuts of meat are rolled to form roasts and tied so that they do not unroll during cooking. **See Figure 4-69.** Stuffed veal breast is commonly tied to retain its shape and hold the stuffing.

Tied Roasts

National Cattlemen's Beef Association

Figure 4-69. Some cuts of meat are rolled to form roasts and tied so that they do not unroll during cooking.

CHECKPOINT 4-13

1. Describe how to trim and cut a beef tenderloin.

2. Trim and cut a beef tenderloin.

3. Describe how to cut a boneless strip into steaks.

4. Cut a boneless strip into steaks.

5. Describe how to french veal chops.

6. French veal chops.

7. Describe four ways to tenderize beef.

8. Tenderize identical cuts of beef four different ways.

9. Describe how to grind fresh meat.

10. Grind fresh beef.

11. Explain the purpose of tying meat.

MARKET FORMS OF PORK

Many breeds of hogs are used in the production of pork products. Some of the more common breeds are the Duroc, Chester White, Hampshire, Yorkshire, and Berkshire. The Berkshire is most desirable because of its intramuscular fat and juiciness. The Berkshire is sometimes called the American Kurobuta because it has many of the same characteristics as the highly prized Japanese Kagoshima Kurobuta.

Pork is available in a variety of market forms, including whole carcasses, primal cuts, fabricated cuts, and variety meats. Knowledge of these market forms is necessary for accurate product ordering.

Whole Carcasses of Pork

Both whole hog and suckling pig carcasses are available. Purchasing a whole hog carcass allows a chef to be more creative with the menu, rather than being restricted to pre-portioned cuts that can be found in most restaurants. A *suckling pig* is a pig 4–6 weeks old that weighs 20–35 lb dressed. Suckling pig carcasses are sold with the head attached and are priced per pig rather than by the pound.

Primal and Fabricated Cuts of Pork

A primal cut is a large cut from a whole or a partial carcass. A hog carcass has two sets of primal cuts. One set is on the left side and one set is on the right side. Each primal cut is divided into fabricated cuts. The five primal cuts of pork are the picnic shoulder, shoulder butt, loin, leg, and belly. **See Figure 4-70.** Some operations have a butcher on staff and order only primal cuts. Other operations order only fabricated cuts.

A fabricated cut is a ready-to-cook cut that is packaged to certain size and weight specifications. Fabricated cuts of pork include hams, roasts, pork chops, pork cutlets, stew meat, and ground pork. Fabricated cuts are a convenient way of providing uniform portions while reducing labor costs. However, the price per pound for fabricated cuts is higher than the price per pound for primal cuts.

Picnic Shoulders. The *picnic shoulder* is a primal cut of pork that is the lower half of the shoulder of a hog. The average weight of a picnic shoulder is approximately 9% of the total carcass weight. Picnics are fabricated from the picnic shoulder primal cut. **See Figure 4-71.** A *picnic* is a cut of pork fabricated from the upper part of the foreleg that includes a portion of the shoulder. A picnic resembles a ham in shape, but is smaller and contains more bone and less lean meat. Fresh picnic can be used to prepare chop suey, pork patties, or pork sausage. Pulled pork is prepared from smoked picnic meat.

Market Forms of Pork

Primal Cuts of Pork

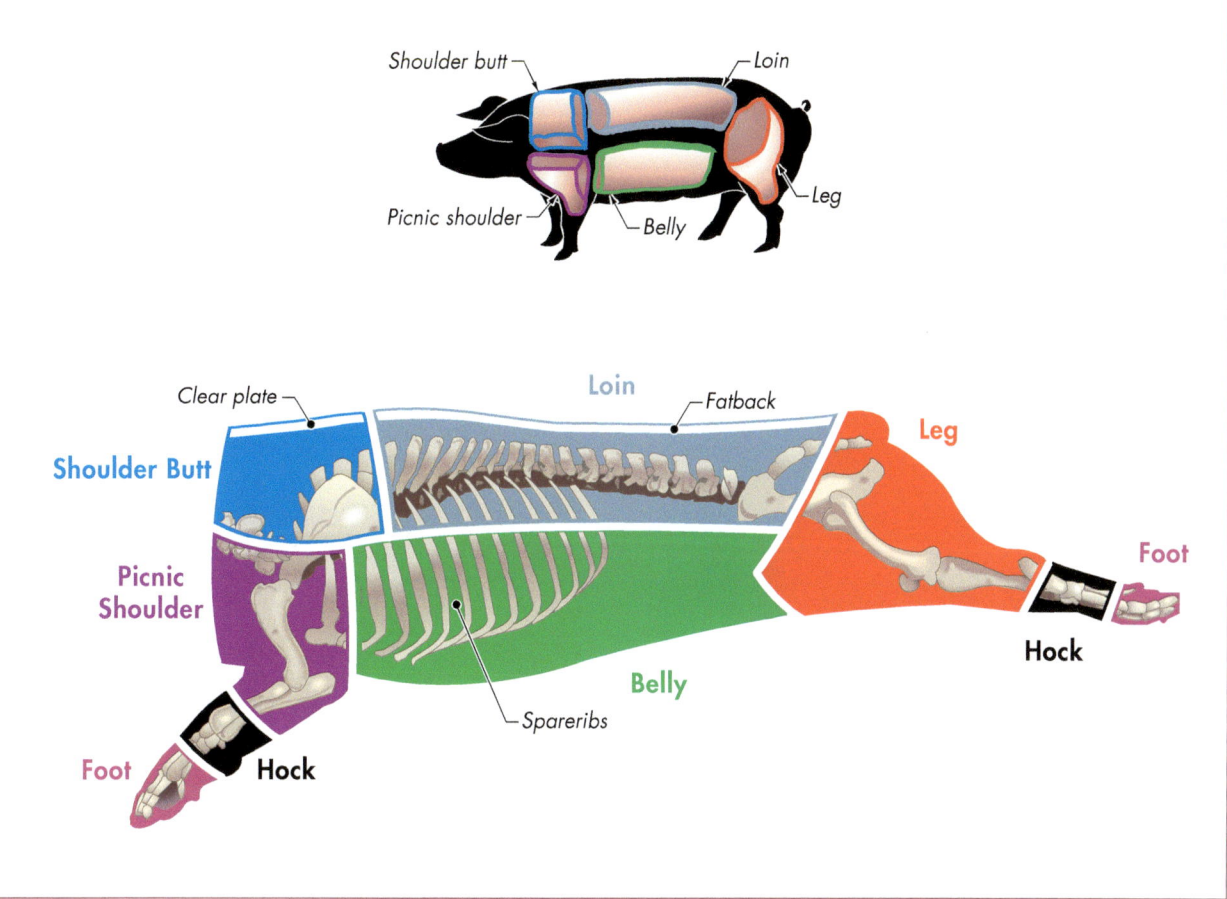

Figure 4-70. The primal cuts of pork include the picnic shoulder, shoulder butt, loin, leg, and belly.

Picnic Shoulder Cut of Pork

Figure 4-71. Picnic shoulders are fabricated from the picnic shoulder primal cut.

Shoulder Butts. The *pork shoulder butt,* also known as Boston butt, is a square, compact area of the shoulder located just above the front legs of a hog. Its average weight is approximately 8% of the total carcass weight. The shoulder butt contains the blade bone and a large portion of lean meat. It is usually sold fresh with the bone. The meat is moderately tough due to the amount of connective tissue, so it is typically roasted or braised. A shoulder butt can also be fabricated into blade steaks, cottage ham, or ground meat. **See Figure 4-72.** A *cottage ham* is the smoked, boneless meat extracted from the blade section of the shoulder butt. Boneless shoulder butts are often tied with string because they fall apart easily when cooked.

The *clear plate* is a rectangular slab of fat that contains a few strips of lean meat located just above the shoulder butt. Clear plate that has been cured in salt is called salt pork. It is often used as a flavoring ingredient in dishes such as beans and bitter greens. Salt pork is often blanched to extract excess salt before it is used as a flavoring ingredient.

Shoulder Butt Cuts of Pork

Boston Butt   **Cottage Ham**   **Blade Steak**

Ground Meat   **Sausages**

Figure 4-72. A shoulder butt can also be fabricated into blade steaks, cottage ham, or ground meat.

Loins. The *pork loin* is a primal cut that extends along the backbone, from about the second rib, through the rib and loin area of a hog. Its average weight is approximately 18% of the total carcass weight, and it can be cut into a variety of fabricated cuts. **See Figure 4-73.** The *pork tenderloin* is a fairly long, tapered strip of lean meat taken from the underside of the loin. Tenderloin is the most tender pork cut and can be prepared using any cooking method.

Baby back ribs are the meaty bones on the rib end of the pork loin. These meaty ribs are only 3–6 inches long and are curved where they meet the backbone. A full slab of baby back ribs has 11–13 ribs. A standing rib roast is the whole pork loin muscle with the baby back ribs attached. A boneless loin and pork cutlets are also fabricated cuts from the loin.

Canadian bacon is the trimmed, pressed, and smoked boneless loin of pork. *Fatback* is the layer of fat that runs along the back of the hog. It can be used to flavor dishes such

as beans and collard greens or added to sausage or ground pork. Lard is usually rendered from fatback.

Legs. A *pork leg* is a primal cut of pork that is composed of the hind thigh and buttock of a hog. A pork leg is sometimes referred to as a ham because it is the cut that is fabricated from the leg. It contains a high proportion of lean meat and the average weight is approximately 24% of the total carcass weight. The ham is a typically cut from the middle of the shank bone to the aitchbone, or hip bone. **See Figure 4-74.**

Unprocessed ham is called fresh ham or green ham. Hams are sold boneless, bone-in, and partially boned. The most popular form of ham is cured in a solution of salt, sodium nitrite, and sugar and then smoked. The skin may be left on a ham or it may be removed.

Ham is often cut into steaks or cutlets and broiled or pan-fried. Ham is also sliced and used to make sandwiches. *Prosciutto* is a type of dry-cured Italian ham. Prosciutto is typically sliced paper thin and used to make hors d'oeuvres or appetizers.

Loin Cuts of Pork

Frenched Loin

Bone-In Loin

Tenderloin

Boneless Loin

Bone-In Chop

Loin Steak

Butterfly Chop

Boneless Chop

National Pork Producers Council

Figure 4-73. A variety of cuts are fabricated from the loin primal cut.

Leg Cuts of Pork

Fresh Ham

Cured Ham

Boneless Leg

Cutlets

National Pork Producers Council

Figure 4-74. A primal leg cut is fabricated into several cuts.

Belly. A *belly* is a primal cut of pork that is the lower portion of the hog between the shoulder and the leg. Its average weight is approximately 19% of the total carcass weight. Spareribs, pork belly, and bacon are fabricated cuts from the belly primal cut. **See Figure 4-75.**

Pork spareribs are the long, narrow ribs and breastbone of a hog. They are quite fatty, yet the meat is tender and has an excellent flavor. A full rack contains 14 ribs. Spareribs may be purchased fresh or smoked. Fresh spareribs should be cooked slowly over low heat until the meat is tender and the fat is rendered. Spareribs are typically seasoned prior to cooking with a rub or a marinade. Cooked spareribs can be browned on the grill or under a broiler to caramelize the meat. They are commonly barbequed, broiled, or roasted and served as an appetizer or an entrée.

Pork belly is a fatty slab of meat and skin from the side and belly of a hog. Braised pork belly is a popular dish. Pork belly can also be cured and served as bacon or pancetta. *Bacon* is pork belly that has been cured and usually smoked. There are three basic cuts of bacon: thin, regular, and thick. Thin bacon is sliced into 22–26 strips per lb. Regular bacon is cut into 16–20 strips per lb, and thick bacon is cut into 12–16 strips per lb. Bacon is used to make entrées, appetizers, and sandwiches. It also is used to garnish soups. Bacon and the fat rendered from bacon are often used to season other foods.

Pancetta, also known as Italian bacon, is unsmoked pork belly that has been cured in salt and spices, such as nutmeg and pepper, and then dried for a few months. Pancetta is typically cut into paper-thin slices before being added to a dish. It adds a distinctive pork flavor, especially to pastas, that is not smoky like bacon.

Pork Offals

Unlike most domesticated meat animals, just about every part of a hog is used for consumption. An offal is an edible part of an animal that is not part of a primal cut. Offals are often ground and used to prepare sausage, meatballs, and fillings. Pork that is to be ground should be very cold before grinding, as should all parts of the grinder that come in contact with the pork. Pork that is cool instead of cold has a tendency to lose its texture and can appear puréed instead of ground. Some chefs place pork in the freezer for a short time before grinding it for better results.

In addition to common offals such as the heart, liver, and kidneys, pork variety meats include jowls, hocks, pig feet, and headcheese. A *jowl* is meat from the cheek of a hog. Jowls are cured and smoked in the same manner as bacon. Jowls are also used to flavor items such as baked beans or long beans.

Hocks are cut from the lower part of the front and hind legs of a hog. They have very little meat, but good flavor, with a good amount of fat, bone, and gristle. Hocks are available fresh, cured, or cured and smoked. They are often used to flavor soups or served with bean dishes.

Belly Cuts of Pork

National Pork Producers Council

Figure 4-75. Spareribs, pork belly, and bacon are fabricated cuts from the belly primal cut.

Pig feet, also known as trotters, can be purchased fresh, smoked, or pickled. Fresh pig feet are commonly broiled or served boneless on a cold plate.

Headcheese is the spiced, pressed, and jellied meat from the head of a hog. The tongue, heart, feet, and quality trimmings may also be ground and included in headcheese. It is often served cold like lunchmeat or used as a binding agent in stocks.

Receiving and Storing Pork

Like all meats, pork is a potentially hazardous food. The color, odor, texture, and temperature must be checked when received. The meat should be pink in color and the fat should be white. There should be no odor, and the texture should be firm and not dry or slick. Pork that is in the temperature danger zone should be rejected. Refrigerated pork should maintain an internal temperature of 41°F or below.

Frozen pork should be stored at temperatures below 0°F. Frozen pork that needs to be thawed should be placed in the refrigerator overnight. Larger cuts of pork may take more than one day to thaw under refrigeration. If necessary, pork can be thawed under running water if it is cooked immediately after thawing. Some frozen cuts, such as sausage patties or breaded pork tenderloins, can be thawed as part of the cooking process.

Vacuum-packed pork must not be opened until needed for service or preparation. The airtight plastic of vacuum packaging helps to preserve meat for three to four weeks. **See Figure 4-76.** Once the vacuum seal is broken, the meat has a shelf life of only two or three days. Cut pork that is not vacuum-sealed should be wrapped tightly, refrigerated immediately, and used as soon as possible.

Vacuum-Packaged Pork

Figure 4-76. Airtight plastic of vacuum packaging helps to preserve meat for three to four weeks.

In addition to packaging, pork can also be irradiated to reduce the risk of potentially harmful microorganisms. The irradiation process does not cook pork and does not have an adverse effect on its appearance or taste. The USDA requires all irradiated food be labeled with an irradiation symbol.

Inspection and Grades of Pork

All pork used in foodservice operations must be procured from USDA inspected plants. At the time of slaughter, a hog carcass is stamped with the round USDA inspection stamp, indicating that the hog was slaughtered at an inspected plant. This stamp does not indicate anything about the quality of the meat. **See Figure 4-77.** The purple inspection stamp is used for whole carcasses and all fabricated and processed meats. It is found either on the meat itself or on the case in which it is packed. The number on the stamp identifies the plant where the animal was processed.

USDA Inspection Stamps

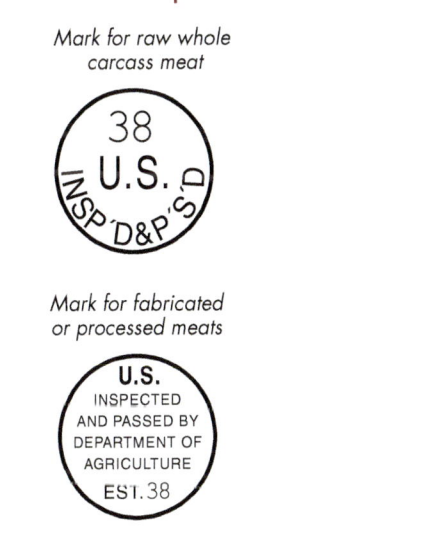

Figure 4-77. A round USDA inspection stamp indicates that a hog was slaughtered at a USDA-inspected plant.

Pork is not graded like beef, veal, and lamb. It is produced from young hogs that were bred and fed to produce uniformly tender meat. Quality pork has very little fat covering on the surface. The meat is firm and grayish-pink in color.

CHECKPOINT 4-14

1. Identify common market forms of pork.

2. Explain the advantage of purchasing a whole carcass.

3. Identify the five primal cuts of pork.

4. Identify the cut fabricated from a picnic shoulder.

5. Identify cuts fabricated from a shoulder butt.

6. Identify cuts fabricated from a pork loin.

7. Identify cuts fabricated from a leg of pork.

8. Identify cuts fabricated from a pork belly.

9. Describe four offals that are only fabricated from pork.

10. Identify four traits that should be checked upon receiving pork.

11. Identify the required storage temperature for refrigerated pork and for frozen pork.

12. Explain why vacuum-sealed packages of pork should only be opened at the time of use.

13. Describe the effects of irradiation on pork.

14. Describe the USDA inspection and grading of pork.

FABRICATING PORK

The size of storage facilities and availability of staff with fabrication skills often determine whether a foodservice operation purchases primal cuts, fabricated cuts, or a combination. Some operations purchase whole hog carcasses and fabricate all the cuts in-house, while others only fabricate the loin. Pork fabrication requires the removal, trimming, and tying of meat as well as butterfly cuts.

Removing Pork Tenderloins

Some food service operations purchase primal loins and fabricate cuts in-house. If a loin is being fabricated in-house, the tenderloin must first be removed from the loin.

Procedure for Removing the Tenderloin from a Loin

1. Trim the surface fat of the loin to approximately ¼ inch thick on both sides.
2. Slide the knife under the exposed rib and slice through to the backbone to separate the tenderloin muscle from the vertebrae.
3. Carefully cut down the center of the backbone between the muscle and the exposed vertebrae.
4. Continue cutting to the end of the vertebrae until the tenderloin is completely removed from the loin.

Once the tenderloin has been removed from the loin, it needs to be trimmed. Trimming a tenderloin involves removing all the fat and silverskin. **See Figure 4-78.**

Pork Tenderloin

Photo Courtesy of D'Artagnan, Photography by Doug Adams Studio

Figure 4-78. Trimming a tenderloin involves removing all the fat and silverskin.

Procedure for Trimming a Tenderloin

1. With a rigid boning knife, carefully remove the chain muscle from the side of the tenderloin and reserve.
2. Use a rigid boning knife to pull back and trim the thick fat covering from the tenderloin.
3. Position the tenderloin so that the head (wide end) is near the cutting hand and the tail (narrow end) is near the guiding hand. Insert the tip of the boning knife just beneath the silverskin at the tail of the tenderloin.
4. Angle the knife blade upward slightly and draw it along the length of the tenderloin, just beneath the silverskin, toward the head. *Note:* Use the guiding hand to hold the silverskin firmly while cutting.

Tying Boneless Pork Roasts

A boneless pork roast falls apart when cooked. To prevent this, the roast is often rolled and then tied with butcher's twine prior to cooking.

Fabricating Boneless Pork Loin
Media Clip

Procedure for Tying a Boneless Roast

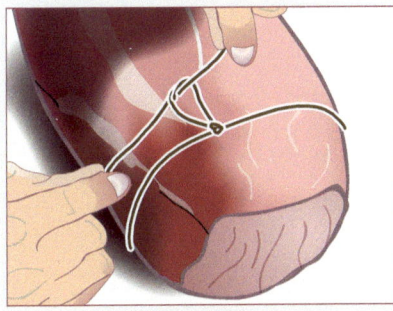

1. Place the outside of the meat facedown and roll the sides tightly toward the center.

2. Wrap butcher's twine around one end of the roast using a slipknot to tighten firmly. Then tie a square knot to secure the tie.

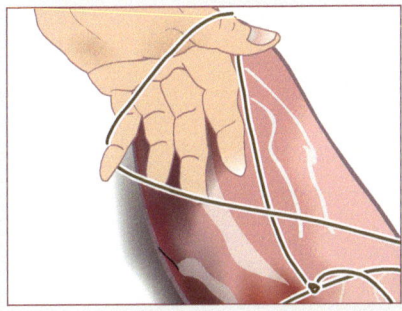

3. Wrap the twine around one hand to form a loop. Then twist the hand over the meat until the palm is facing down.

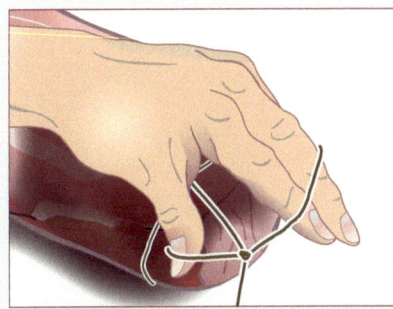

4. Move the palm to the front of the roast and slide the loop over the meat.

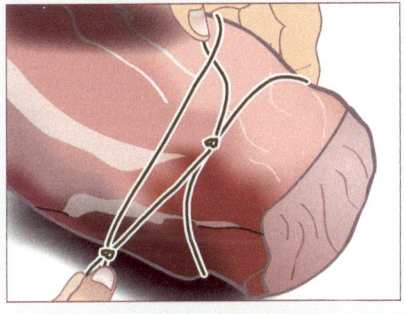

5. Remove the hand from the loop and secure the twine around the meat.

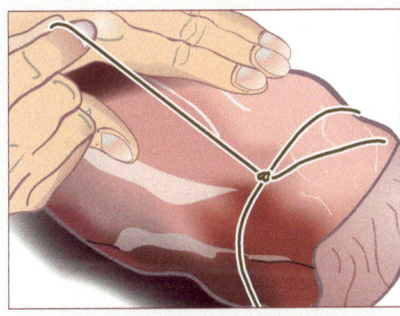

6. Pull tightly to straighten the loop.

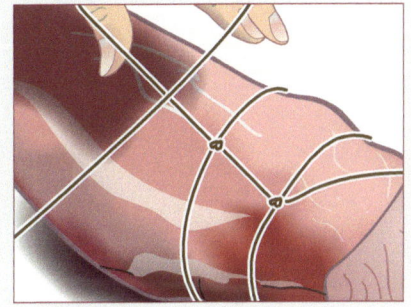

7. Continue making loops in the same fashion 1½ inches apart down the length of the roast. *Note:* Check each loop to make sure it is an even distance from the last loop.

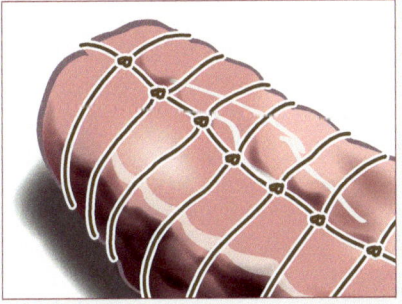

8. Turn the roast over. Then pull the twine over the first loop and tuck it under and around the loop, pulling it toward the front of the roast.

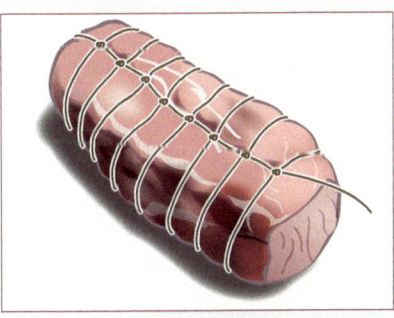

9. Continue pulling over the loops until the underside of the roast is fully tied.

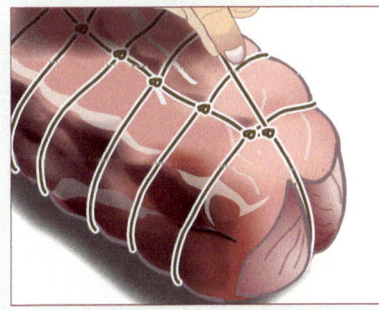

10. Turn the roast over and bring the twine up to tie a square knot to secure the roast. Cut the twine close to the knot.

Tying a Pork Loin
Media Clip

Making Butterfly Cuts

Thicker cuts such as pork chops can be butterfly cut to make a pocket for stuffing or to create a thinner product that will cook more quickly. To make a butterfly cut, the chop is cut almost completely in half horizontally.

Procedure for Making a Butterfly Cut

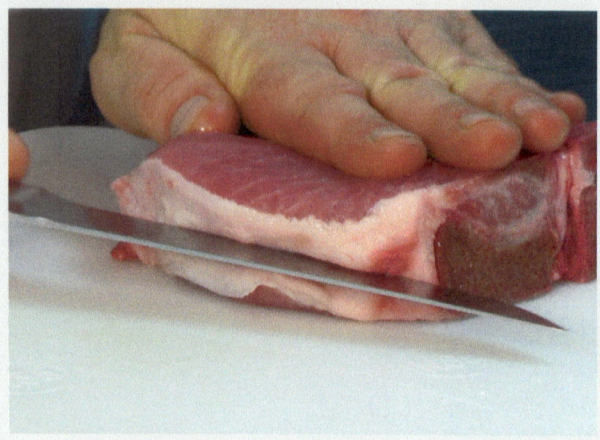

1. Place the pork chop near the edge of a cutting board to prevent the knuckles of the knife hand from hitting the board.
2. Place the guiding hand flat on top of the chop to hold it in place. While holding the knife blade parallel to the cutting board, place the blade edge at the midpoint of the chop.

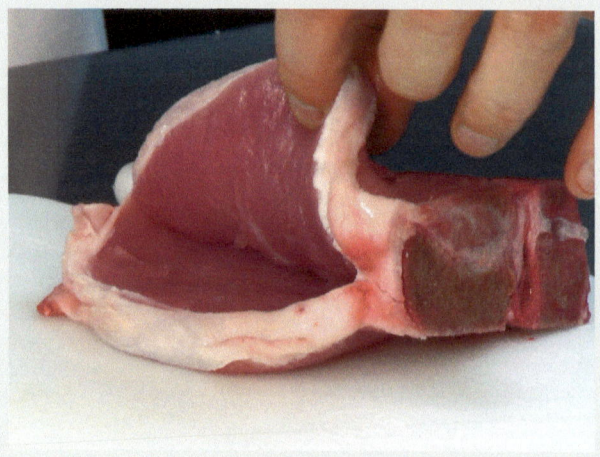

3. Slice the chop almost all the way through, leaving ¼–½ inch connected at the back side.

CHECKPOINT 4-15

1. Describe how to remove the tenderloin from a pork loin.
2. Remove the tenderloin from a pork loin.
3. Describe how to trim a pork tenderloin.
4. Trim a pork tenderloin.
5. Describe how to tie a boneless pork roast.
6. Tie a boneless pork roast.
7. Describe how to butterfly boneless pork chops.
8. Butterfly a boneless pork chop.

MARKET FORMS OF LAMB

Lamb is the meat from slaughtered sheep that are less than a year old. It is important to understand the basic composition and bone structure of a lamb before working with the meat. Lamb meat has smooth grain and is similar in color to beef. It is also very tender and has a mild flavor.

American lamb is weaned to grain, then hay, and finished with a formulated feed that contains wheat, sorghum, and vitamins. American lamb has a light gamey flavor. Australian and New Zealand lambs are smaller in size and have a more pronounced flavor than American lamb. The pronounced flavor has been attributed to their grass diet.

Lamb is available in a variety of market forms, including whole and partial carcasses, primal cuts, fabricated cuts, and offals. Knowledge of these different market forms is necessary for accurate product ordering.

Whole and Partial Lamb Carcasses

Both whole and partial lamb carcasses are available, but partial carcasses are not commonly purchased by foodservice operations. A partial carcass is a primary division of a whole carcass. Partial carcasses of lamb include foresaddles, hindsaddles, backs, and bracelets. Like whole carcasses, partial carcasses of lamb are often not purchased due to the skilled labor and storage space required to process them. Whole and partial carcasses also yield cuts of lamb that may not be used by the foodservice operation and are therefore wasted.

Foresaddle and Hindsaddle. Lamb is typically split into head and tail sections known as the foresaddle and the hindsaddle. The foresaddle and hindsaddle are split between the 12th and 13th ribs, not down the backbone. The *lamb foresaddle* is the front half of the carcass consisting of the primal shoulder, rack, and breast cuts. The *lamb hindsaddle* is the rear half of the carcass consisting of the loin and leg.

Back and Bracelet. Lamb carcasses are also divided into backs and bracelets. A *lamb back* is a lamb rack and loin that are still joined. A back is often purchased by foodservice operations that sell a lot of lamb chops because chops can be cut from the end of the rack to the end of the loin.

A *lamb bracelet* is a hotel rack with the breast still attached. **See Figure 4-79.** A *lamb hotel rack* is a lamb back that remains joined along the backbone. A bracelet is typically fabricated

into a hotel rack and a breast. Leaving the breast attached to the rack yields a larger amount of meat and fat.

Primal and Fabricated Cuts of Lamb
Unlike beef, lamb is not split into sides before being divided into primal cuts. The left and right primal cuts remain joined together and are purchased as a single cut. For example, the leg primal cut has two joined legs. The five primal cuts of lamb are the shoulder, rack, loin, leg, and breast/shank. **See Figure 4-80.**

Lamb Bracelet

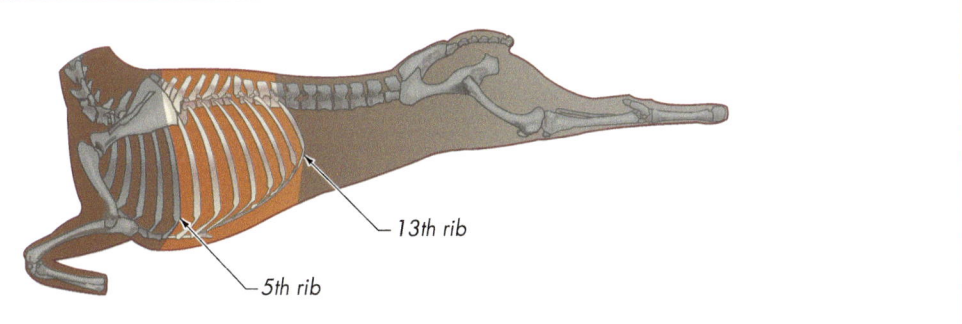

Figure 4-79. A lamb bracelet is a hotel rack with the breast still attached.

Primal Cuts of Lamb

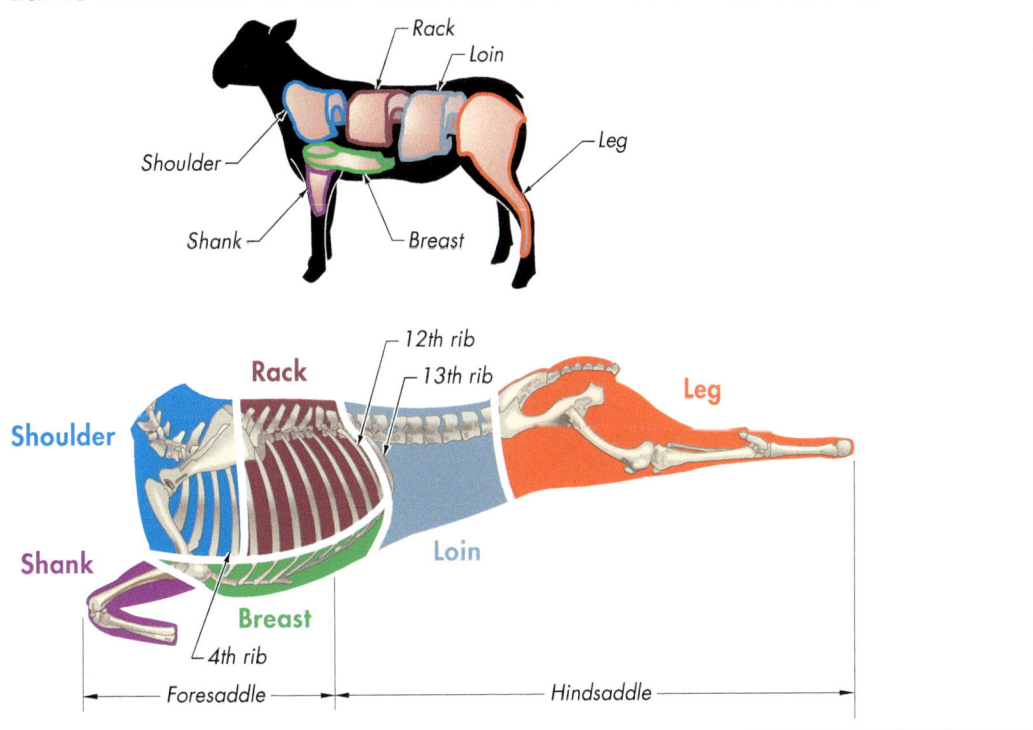

Figure 4-80. The primal cuts of lamb are the shoulder, rack, loin, leg, and breast/shank.

Each primal cut is further divided into fabricated cuts. Some fabricated cuts of lamb include stew meat, ground meat, hotel racks, roasts, chops, and leg of lamb.

Fabricated cuts are a convenient way of providing uniform portions while reducing labor costs. The price per pound for fabricated cuts is higher than the price per pound for primal cuts. Some operations order only primal cuts. Other operations order only fabricated cuts. Most foodservice operations primarily purchase some primal cuts and some fabricated cuts of lamb.

To help ensure that foodservice operations and suppliers communicate efficiently, the USDA publishes the Institutional Meat Purchase Specifications (IMPS) for commonly purchased meats and meat products. Cuts of lamb are numbered by category.

Shoulder. A *lamb shoulder* includes the first four rib bones of each side and the arm and neck bones. Its average weight is about 36% of the total carcass weight. Shoulder meat is quite lean and has excellent flavor. It is seldom cooked whole because of its many small bones and connective tissues. Instead, it is fabricated into a variety of cuts, including roasts, chops, stew meat, and ground meat **See Figure 4-81.** Ground lamb meat can be used in the preparation of sausages, meatballs, and meat fillings.

Rack. A *lamb rack* is eight rib bones located between the shoulder and loin of a lamb. Its average weight is about 8% of the total carcass weight. The meat is tender and well-marbled because it comes from an area of the back where the muscles are not worked much. A rack is often split along the backbone into two racks. A *lamb crown roast* is a hotel rack containing sixteen ribs with the bones frenched, notched, and tied to create a circle to resemble a crown. Paper frills are often used to cover the exposed tip of each rib bone of a crown roast. *Frenching* is a method of removing the meat and fat from the end of a bone and is generally applied to chops. **See Figure 4-82.**

A rack of lamb is sometimes coated with herbs, roasted whole, and then sliced to order. However, in most cases it is cut into single or double chops and grilled or broiled. A *double rib lamb chop* is a rib chop cut to a thickness equal to two standard rib chops. An *English lamb chop* is a 2 inch thick fabricated cut taken along the entire length of the unsplit loin. Chops may also be frenched.

Loin. A *lamb loin* is a primal cut located between the rack and leg that includes the 13th rib, the loin eye muscle, the center section of the tenderloin, the strip loin, and some flank meat. **See Figure 4-83.** An unsplit primal lamb loin is commonly known as a saddle and has an average weight of about 13% of the total carcass weight.

Shoulder Cuts of Lamb

American Lamb Board

Figure 4-81. The shoulder is often fabricated into a variety of cuts, including roasts, chops, stew meat, and ground meat.

Rack Cuts of Lamb

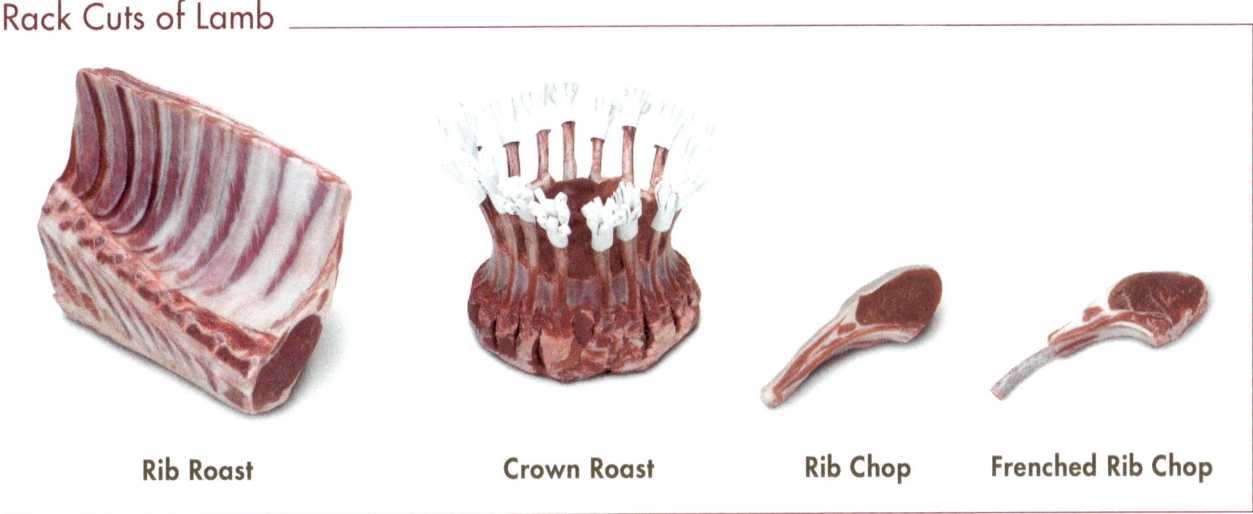

Rib Roast   **Crown Roast**   **Rib Chop**   **Frenched Rib Chop**

Figure 4-82. A rack is fabricated into rib roasts, crown roasts, and rib chops.

Loin Cuts of Lamb

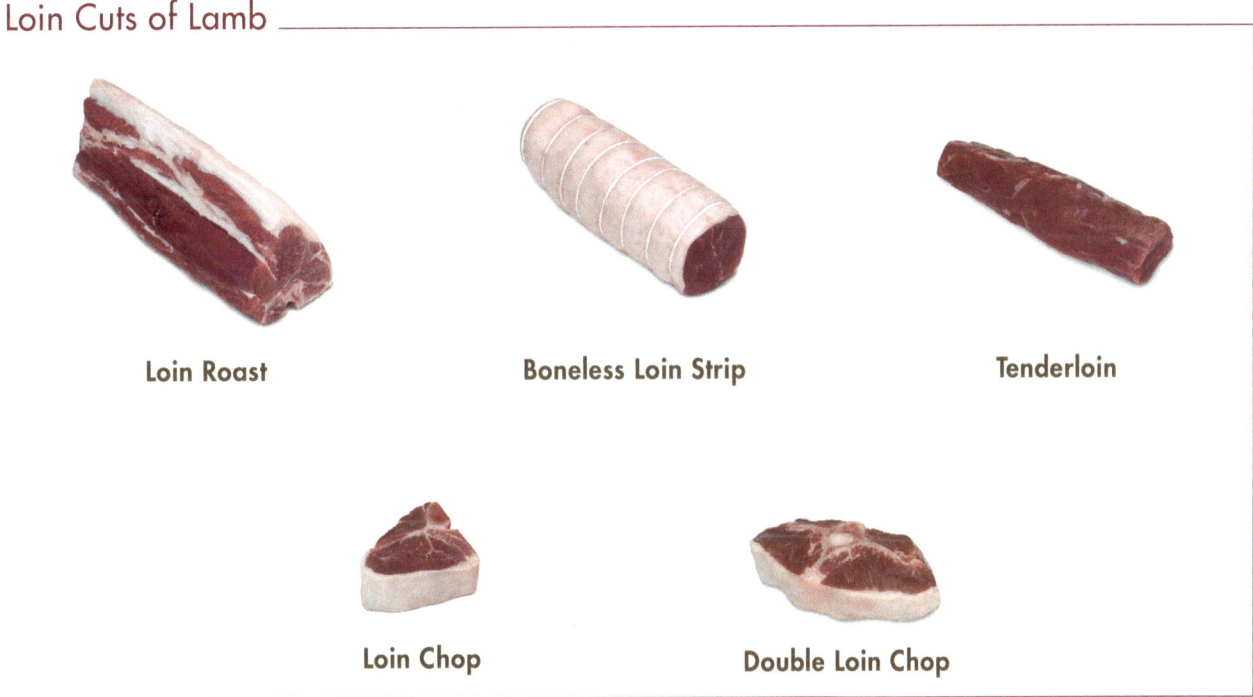

Loin Roast   **Boneless Loin Strip**   **Tenderloin**

Loin Chop   **Double Loin Chop**

Figure 4-83. A primal loin can be fabricated into a loin roast, boneless loin strip, tenderloin, loin chop, and double loin chop.

Fabricated cuts from the loin are best prepared using dry-heat cooking methods such as grilling, broiling, or roasting. Lamb loins are typically cut into boneless or bone-in chops that can be grilled or broiled. The tenderloin can be removed and cut into noisettes or can be roasted whole. A *noisette* is a small, round, boneless medallion of meat.

Leg. Lamb legs are not split into two separate legs. Lamb legs remain joined at the hip. A *leg of lamb* is a primal cut of lamb that contains the last portion of the backbone, hip bone, aitchbone, round bone, hindshank, and tailbone. The aitchbone is the buttock or rump bone and is located at the top of the leg. The leg accounts for approximately 34% of the total carcass weight. Leg of lamb includes part of the sirloin, the top round, bottom round, and knuckle meat. **See Figure 4-84.**

The leg contains lean, fine-textured meat that is more tender near the sirloin end and tougher toward the shank end. Lamb leg is commonly split in two and partially boned, stuffed, and roasted. It can also be split in two and completely boned, rolled, tied, and roasted. Single lamb legs can be purchased boned, rolled, and tied (BRT). Meat from the sirloin end can also be cut into lamb steaks. Meat from the shank end is most commonly used for stew meat or ground for patties. Shank meat is commonly cut in cross-sections, braised, and served in a rich, flavorful sauce.

Breast/Shank. A *lamb breast* is a thin, flat, primal cut of lamb that contains the breastbone, the tips of the rib bones, and cartilage that is located under the shoulder and ribs. The breastbone is actually cartilage because the animal is so young. The breast also includes the shank and weighs about 17% of the total carcass weight. **See Figure 4-85.** A *lamb shank* is a cut of lamb that contains the upper foreshank bones. The small section of seven or more rib tips can also be braised and is often referred to as Denver ribs or lamb riblets. A *lamb riblet* is a rectangular strip of meat cut from the lamb breast that contains part of a rib bone.

Leg Cuts of Lamb

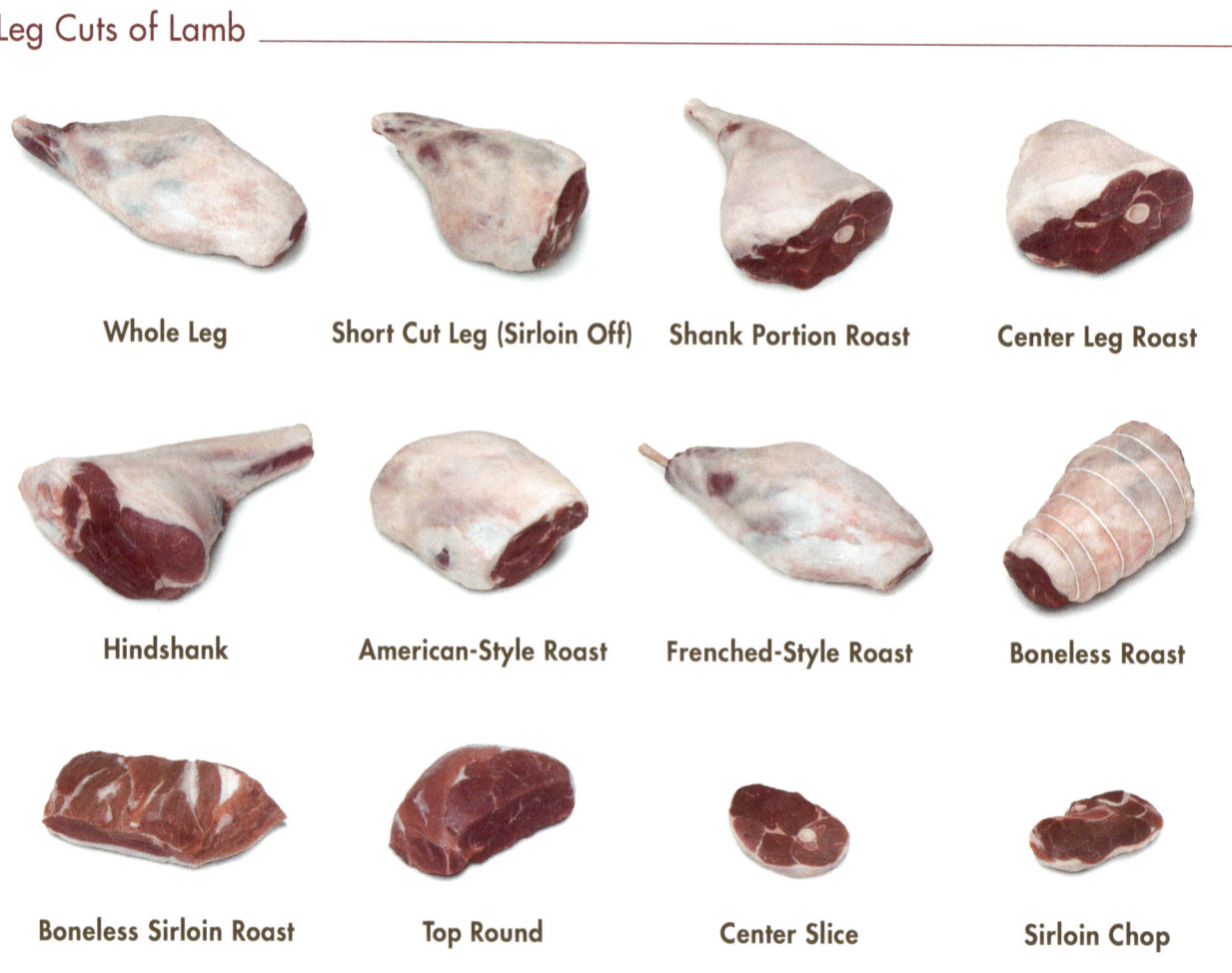

Whole Leg   Short Cut Leg (Sirloin Off)   Shank Portion Roast   Center Leg Roast

Hindshank   American-Style Roast   Frenched-Style Roast   Boneless Roast

Boneless Sirloin Roast   Top Round   Center Slice   Sirloin Chop

American Lamb Board

Figure 4-84. A primal leg of lamb is fabricated into many cuts.

Breast/Shank Cuts of Lamb

Foreshank   Spareribs   Riblets

American Lamb Board

Figure 4-85. A primal breast/shank is fabricated into a foreshank, spareribs, and riblets.

The breast is not a popular cut of meat, but the shape and ample connective tissue make the breast a good cut to stuff, roll, tie, and then braise. Braising breaks down the connective tissue and yields a very tender rolled roast.

Lamb Offals

Lamb offals are not commonly used in most foodservice operations. However, lamb tongue, kidneys, liver, heart, and sweetbreads are featured dishes in Mediterranean and Middle Eastern cuisine. Lamb intestines may also be used as casings for small sausages.

Receiving and Storing Lamb

Like all meats, lamb is a potentially hazardous food and the color, odor, texture, and temperature must be checked upon receipt. The meat should be light red in color and the fat should be white. **See Figure 4-86.** There should be no odor, and the texture should be firm and not dry or slick. Refrigerated lamb should maintain an internal temperature of 41°F or below. Lamb that is in the temperature danger zone should be rejected.

Frozen lamb should be stored at temperatures below 0°F. Frozen lamb that needs to be thawed should be placed in the refrigerator overnight. Larger cuts of lamb, such as legs, may take more than one day to thaw under refrigeration.

Vacuum-packed lamb must not be opened until needed for service or preparation. Once the vacuum seal is broken, the meat has a shelf life of only 2–3 days. Cut lamb that is not vacuum-sealed should be wrapped tightly, refrigerated immediately, and used as soon as possible.

Receiving Standards

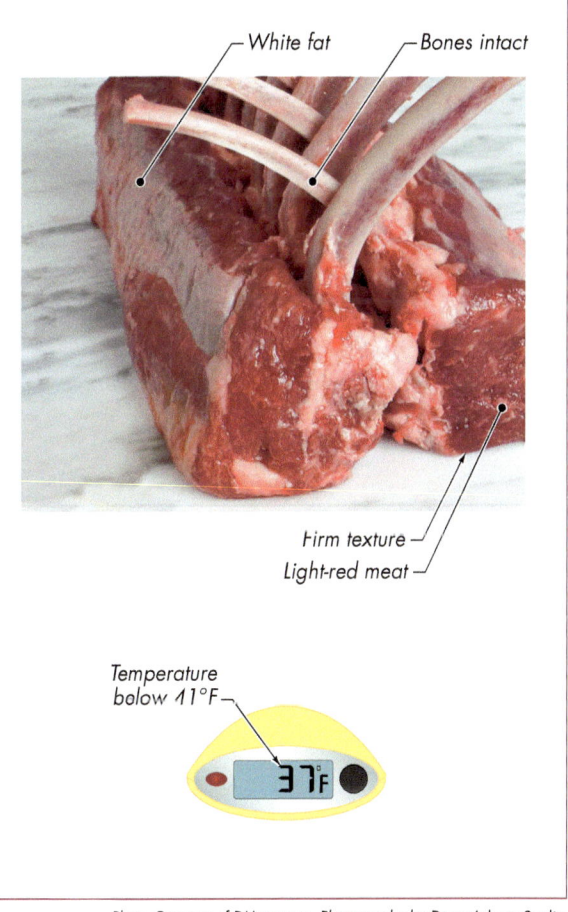

White fat — Bones intact

Firm texture — Light-red meat —

Temperature below 41°F —

37°F

Photo Courtesy of D'Artagnan, Photography by Doug Adams Studio

Figure 4-86. Upon receiving, lamb cuts should be checked for color, odor, texture, and temperature. The bones should be intact.

Inspection and Grades of Lamb

All lamb used in foodservice operations must be purchased from USDA-inspected plants. At the time of slaughter, the lamb carcass or the inspection tag is stamped with the round USDA inspection stamp, indicating the lamb was slaughtered at an inspected plant. This stamp does not indicate anything about the quality of the meat. The inspection stamp is used for whole and partial carcasses as well as all fabricated and processed meats. It is found either on the meat itself, on the tag attached to the meat, or on the case in which it is packed. **See Figure 4-87.** The number on the stamp identifies the plant where the animal was processed.

Quality Grades

Unlike inspection, USDA quality and yield grading is optional for lamb producers. Quality and yield grading stamps are stamped onto carcasses in the same manner as inspection stamps. Quality grading is based on the overall tenderness, juiciness, and flavor of the meat. However, quality grades do not guarantee these characteristics.

The grades of lamb commonly used in foodservice operations are USDA Prime and USDA Choice. **See Figure 4-88.** USDA Prime lamb is well marbled. USDA Choice lamb has slightly less marbling, but is still the most popular grade of lamb used in foodservice operations.

USDA Quality Grade Stamps

Figure 4-88. USDA Prime and USDA Choice lamb is used in foodservice operations.

Yield Grades

Lamb can also be yield graded for the percentage of edible meat to fat and bone. Yield grade shields are numbered 1 to 5 and indicate how much usable meat can be obtained from a lamb carcass. A grade of 1 indicates the highest yield of meat, and a grade of 5 indicates the lowest yield.

USDA Inspection Stamps

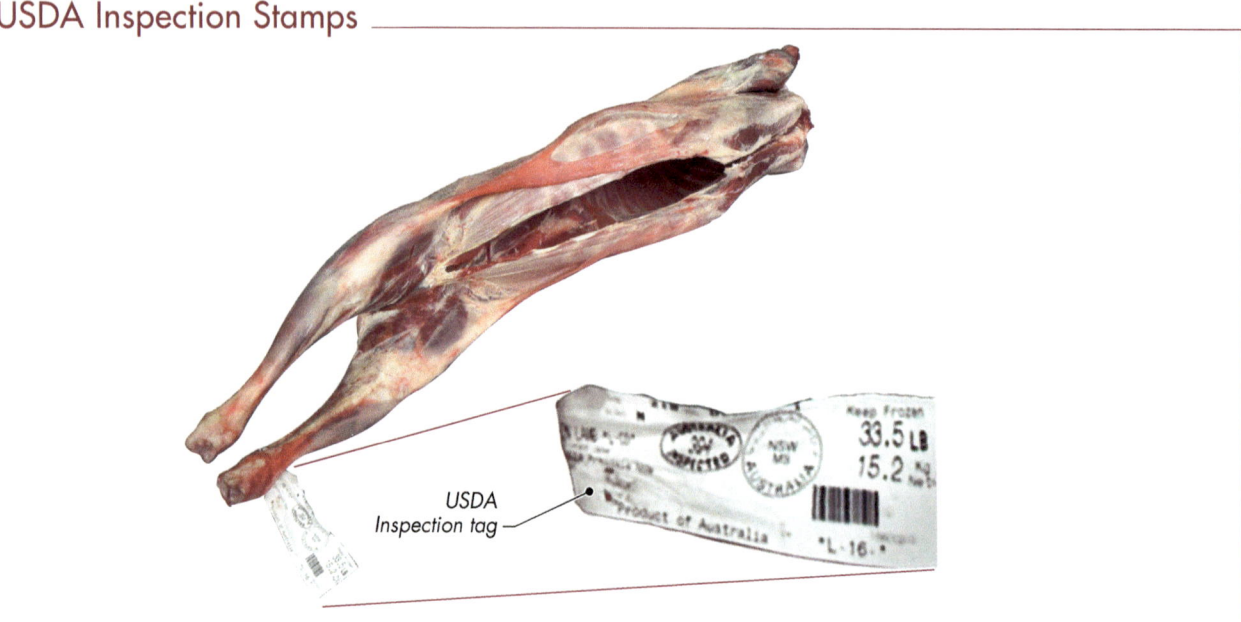

USDA
Inspection tag

Figure 4-87. A USDA inspection stamp is sometimes attached to a lamb carcass in the form of a tag.

CHECKPOINT 4-16

1. Identify common market forms of lamb.

2. Explain the difference between a foresaddle and a hindsaddle.

3. Explain the difference between a back and a bracelet.

4. Identify the five primal cuts of lamb.

5. Identify the cuts fabricated from a lamb shoulder.

6. Identify the cuts fabricated from a lamb rack.

7. Identify the cuts fabricated from a lamb loin.

8. Identify the cuts fabricated from a leg of lamb.

9. Identify the cuts fabricated from a lamb breast.

10. Identify the lamb offals that are used in some cuisines.

11. Identify four traits that should be checked upon receiving lamb.

12. Identify the required storage temperature for refrigerated lamb and frozen lamb.

13. Describe the USDA inspection and grading of lamb.

FABRICATING LAMB

The size of storage facilities and availability of staff with the fabrication skills often determine whether a foodservice operation purchases primal cuts, fabricated cuts, or a combination. Lamb fabrication involves separating and frenching racks, boning and tying loins and legs, and cutting noisettes.

Fabricating Lamb

Separating Hotel Racks

The first step in fabricating a rack is to separate the hotel rack. This yields two separate racks.

Procedure for Separating a Hotel Rack

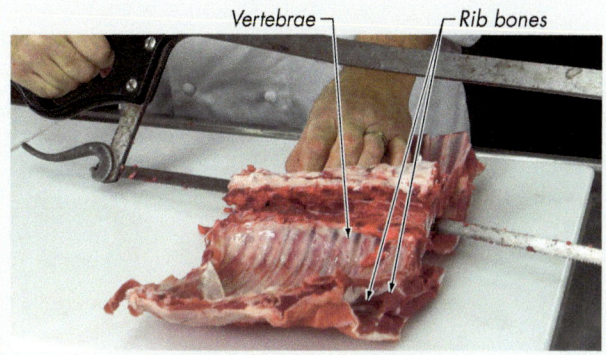
Vertebrae — — Rib bones

1. Turn the rack upside-down, with the ribs pointing upward.
2. Use a meat saw to cut at a 45° angle between the base of the rib bones and the vertebrae.
3. Cut completely through the bones, using caution not to cut the meat.

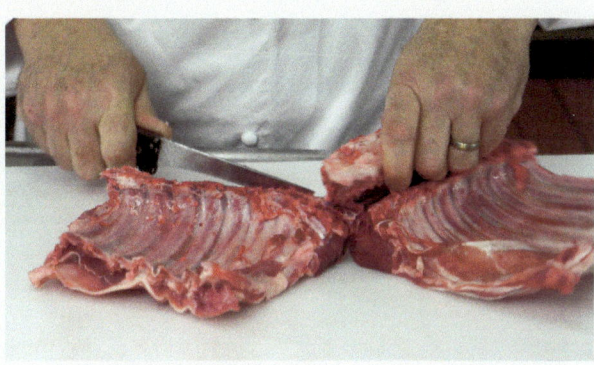

4. Use a knife to finish separating the meat from one side of the vertebrae.
5. Repeat on the other rack.

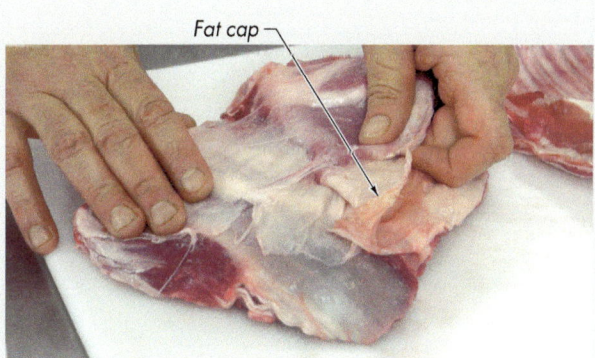
Fat cap —

6. Remove the back strap and fat cap. *Note:* The back strap is a tough tendon that runs parallel to the vertebrae.

Frenching Racks

A rack of lamb is often frenched for service. Several techniques are used to french racks of lamb. Regardless of the technique used to french a rack of lamb, the meat is removed from the end of each bone.

Procedure for Frenching Racks of Lamb

1. Starting on the top side of the rack, score the meat all the way to the bone, 1 inch above the eye meat.
2. Turn the rack over and score the thin membrane on the back of each rib bone from the scored cut to the tip of the bone.

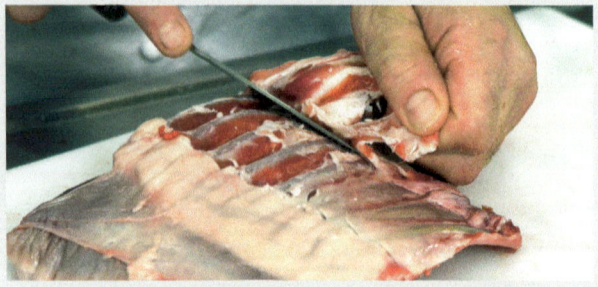

3. From each scored mark, scrape the meat away from the top of the rib to expose the bone.

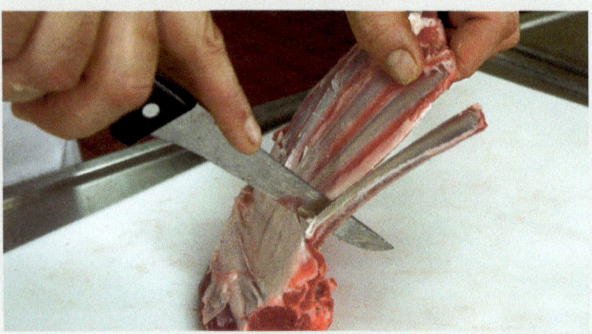

4. Cut between each rib bone down to the scored mark to remove the meat.

Boning and Tying Lamb Loins

A whole lamb loin is often boned and then tied to form a rolled roast. It is important to remove all connective tissue from the skin side of the loin.

Procedure for Boning a Loin to Form a Rolled Roast

1. Trim the layer of connective tissue from the skin side of the loin.
2. Place the skin side down and remove the fat from around the tenderloin.
3. Cut under the vertebrae to separate the bones from the tenderloin. Be careful not to remove the tenderloin. Repeat this step on the other side.
4. Use a knife to cut under the rib bones to the backbone. This cut will free the rib eye from the bones.
5. Remove the backbone by pulling the meat away from the bones and making small cuts as needed.
6. Check the trimmed piece of loin for bone or connective tissue.
7. With the skin side facing up, closely trim the surface so there is about ¼ inch of fat remaining.
8. Place the outside of the meat face down and roll the sides tightly toward the center.

Rolled loin roasts are commonly tied before roasting. A rolled roast is tied to maintain a consistent shape and ensure even cooking.

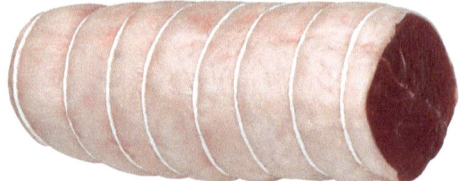

American Lamb Board

Boning and Tying Legs of Lamb

The bone is commonly removed from a leg of lamb before cooking. A boned leg of lamb makes a nice presentation.

A leg of lamb may be tied before roasting. Tying helps the meat maintain a consistent shape and cook evenly during the cooking process.

Frenching Racks of Lamb
Media Clip

Procedure for Boning a Leg of Lamb

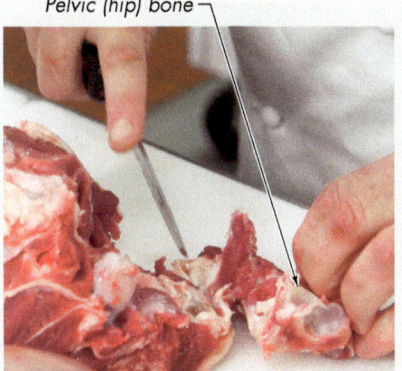
Pelvic (hip) bone

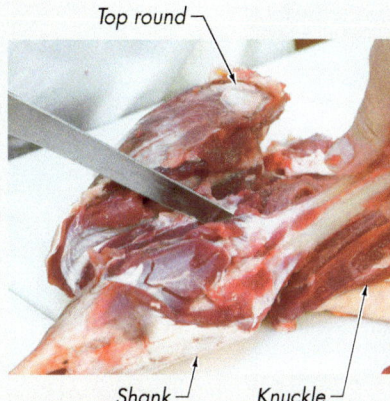
Top round
Shank *Knuckle*

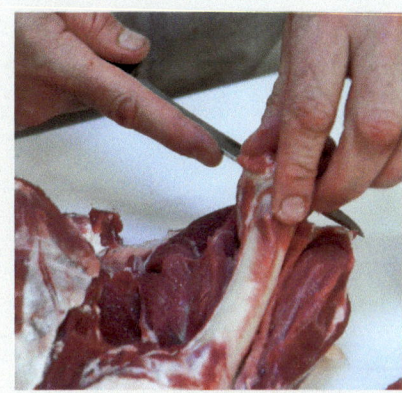

1. Hold the pelvic bone while cutting around the bone to free the pelvic bone from the leg.
2. Carefully remove the top round by cutting lengthwise along the leg bone.

3. Continue trimming around the knee joint to separate the knuckle meat from the bone.

4. Working near the shank end of the leg, continue to cut lengthwise to free the remaining meat from the bone.

Cutting Noisettes

Noisettes are small, round, boneless medallions of meat. A lamb tenderloin may be cut into noisettes. Noisettes may also be cut from a boned loin chop. Once the bone is removed, the loin is rolled and then can be cut into noisettes.

CHECKPOINT 4-17

1. Describe how to separate a hotel rack.
2. Separate a hotel rack.
3. Describe how to french a rack of lamb.
4. French a rack of lamb.
5. Describe how to bone and tie a lamb loin.
6. Bone and tie a lamb loin.
7. Describe how to bone and tie a leg of lamb.
8. Bone and tie a leg of lamb.
9. Describe how to cut noisettes.
10. Cut a lamb tenderloin into noisettes.

FABRICATING RABBIT

Some menus may require the fabrication of specialty game such as rabbit. Rabbits sold for consumption are often crosses between New Zealand and Belgian varieties, Chinese rabbits, or Scottish hares. Rabbits and other specialty game meats must be purchased from insured vendors and stored with care to ensure food safety. Rabbit meat is often sold frozen. Uncooked rabbit is similar in appearance and texture to uncooked chicken. **See Figure 4-89.**

Uncooked Rabbit

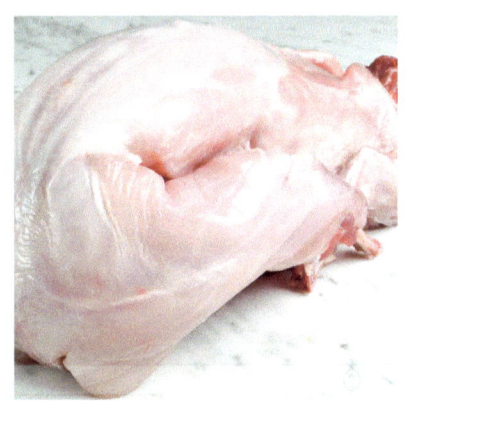

Photo Courtesy of D'Artagnan, Photography by Doug Adams Studio

Figure 4-89. Uncooked rabbit is similar in appearance and texture to uncooked chicken.

Boning a Leg of Lamb
Media Clip

Procedure for Fabricating a Rabbit

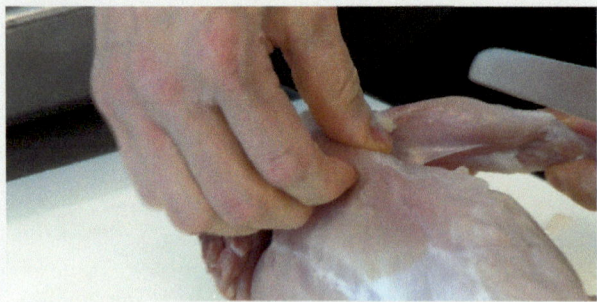

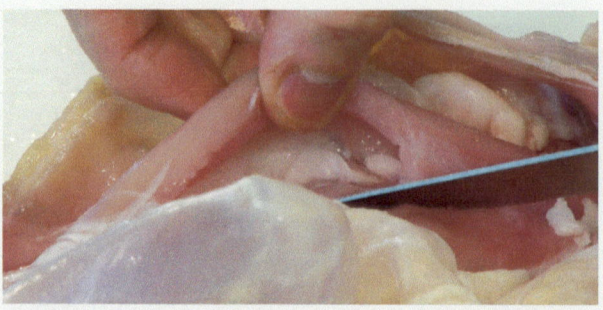

1. Place the rabbit on its side with the forelegs facing the cutting edge of the table. Slice under the joint where one of the forelegs connects to the body. Continue to cut through the joint to remove the foreleg.
2. Repeat the process to remove the other foreleg.
3. Place the rabbit on its back and open the chest cavity.

4. To remove the two tenderloins, insert the tip of the knife under one of the tenderloins and make small cuts in order to remove the meat from the rabbit.
5. Repeat the process to remove the second tenderloin.

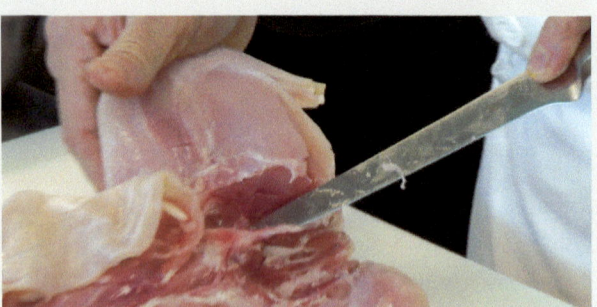

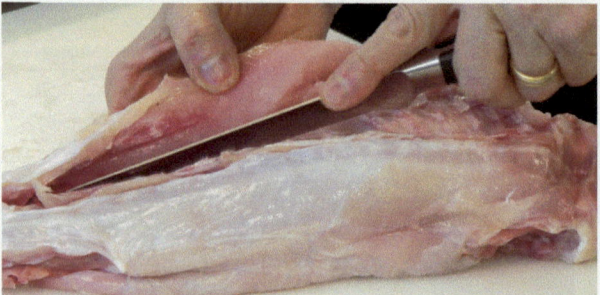

6. To remove the hind legs, firmly hold the back leg and cut at the joint where the leg meets the loin.
7. Peel the meat back to expose the joint and cut through to the backbone. Pull the leg back toward the spine to crack open the joint and cut through the separation.
8. Repeat the process to remove the remaining back leg.

9. To remove the two loins, position the rabbit with the back facing upward. Cut along either side of the spine by making a small incision with the point of the knife and then running the tip along the side of the spine. Repeat this shallow cut, going a little deeper each time.
10. Follow the bone structure of the ribs and continue to cut away the loin muscle from the spine.

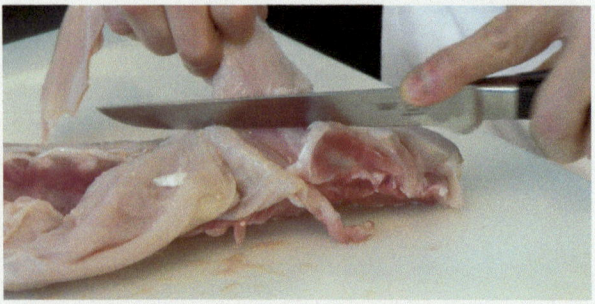

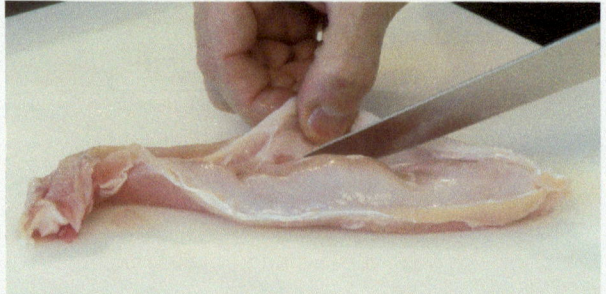

11. To finish separating the loin, cut straight down at the shoulder joint.
12. Repeat the process to remove the loin on the opposite side of the spine.

13. Place one loin silverskin-side down. Make a small cut into the flesh at the thicker end of the loin in order to grasp the silverskin. Holding the silverskin, slice along the silverskin to separate it from the loin.
14. Repeat the process to remove the silverskin from the second loin.

A whole rabbit is fabricated into two forelegs, two tenderloins, two loins, and two hind legs. **See Figure 4-90.** The rabbit carcass and trimmings are often used to make soups and stews.

CHECKPOINT 4-18

1. Describe the appearance and texture of rabbit meat.

2. Identify eight cuts fabricated from a rabbit.

3. Fabricate a rabbit.

Rabbit Fabrication

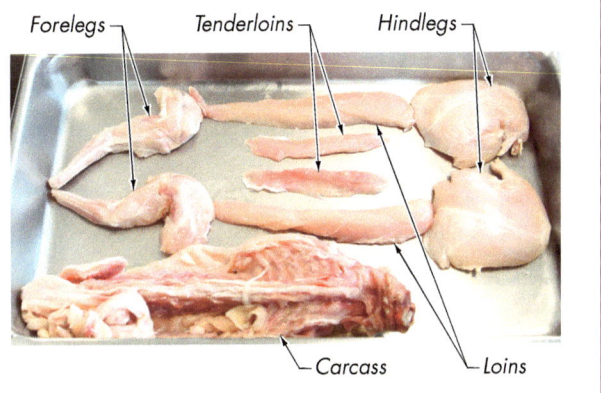

Figure 4-90. A whole rabbit is fabricated into two forelegs, two tenderloins, two loins, and two hind legs.

KEY TERMS

Refer to CD-ROM for **Flash Cards**

- **abalone:** A univalve contained in a brown, bowl-shaped shell with an iridescent, multicolored interior.
- **adductor muscle:** A muscle that opens and closes the shell of a bivalve.
- **airline breast:** A boneless skin-on chicken breast with the first wing section (bone-in) attached.
- **Atlantic oyster:** A variety of oyster that has a fairly flat shell and a distinctive, salty-flavored, plump, and tender flesh. Also known as an Eastern oyster.
- **baby back ribs:** The meaty bones on the rib end of the pork loin.
- **baby T-bone:** A 6–8 oz steak cut from the loin of veal.
- **bacon:** Pork belly that has been cured and usually smoked.
- **bay scallop:** A fairly small scallop harvested from shallow saltwater.
- **beef:** The flesh of domesticated cattle.
- **beef brisket:** A thin section of beef that contains some of the ribs, the breastbone, and layers of lean muscle, fat, and connective tissue.
- **beef chuck:** A shoulder primal cut of beef that contains the first five rib bones, some of the backbone, and a small amount of the arm and blade bones.
- **beef flank:** A primal cut of beef that includes the thin, flat section of the hindquarters located beneath the loin.
- **beef rib:** The primal cut of beef located between the chuck and short loin; contains seven rib bones, from the 6th to the 12th rib.
- **beef round:** A primal beef cut of beef that includes a large grouping of muscles that represent the hind hip and thigh of the carcass.

- **beef rump roast:** A roast cut from the primal round, above the back end of the hip bone.
- **beef shank:** A bony section of beef that is surrounded by a small amount of very tough but flavorful meat.
- **beef short loin:** A primal cut of beef located just to the rear of the primal rib; includes the 13th rib and a small section of the backbone.
- **beef short plate:** A primal cut of beef that includes a thin portion of the beef forequarter located just beneath the rib cut.
- **beef sirloin:** A primal cut of beef situated just behind the short loin; contains some of the backbone and hip bone.
- **beef strip loin:** A short loin without a tenderloin.
- **beef tenderloin:** An eye-shaped muscle running from the primal rib cut into the primal leg.
- **belly:** A primal cut of pork that is the lower portion of the hog between the shoulder and the leg.
- **bivalve:** A mollusk with a top shell and a bottom shell connected by a central hinge that can close for protection.
- **blood line:** A dark-red, almost black muscle that runs along the lateral line of a fish.
- **blue crab:** A North American crab with blue claws and a dark blue-green, oval shell.
- **breast:** The top front portion of the flesh above the rib cage.
- **breast quarter:** Half of a breast, a wing, and a portion of the back.
- **broiler/fryer:** A young male or female chicken that is less than five months old.

KEY TERMS (continued)

- **broiler/fryer duckling:** A duck that is less than two months old.
- **butterflied fillet:** Two single fillets from a dressed fish that are held together by the uncut back or belly of the fish.
- **capon:** A surgically castrated male chicken that is less than four months old.
- **cartilaginous fish:** Any fish that has a skeleton composed of cartilage instead of bones.
- **cassoulet:** A dish that consists of white beans stewed with duck fat, fresh sausage, and duck confit.
- **caul fat:** A meshlike fatty membrane that surrounds sheep or pig intestines.
- **cephalopod:** Any of a variety of mollusks that do not have an external shell.
- **chicken cutlet:** A boneless, skinless section of a chicken breast that has been tenderized.
- **clam:** A bivalve found in both freshwater and saltwater.
- **clear plate:** A rectangular slab of fat that contains a few strips of lean meat located just above the shoulder butt.
- **cockle:** A 1 inch wide bivalve with a shell that has deep ridges.
- **conch:** A univalve that has a pinkish-orange shell with an interior that resembles a large snail.
- **confit:** A French term for meat that has been cooked and preserved in its own fat.
- **Cornish game hen:** Either a female or male chicken that is less than five weeks old. Also known as a Rock Cornish game hen.
- **cottage ham:** The smoked, boneless meat extracted from the blade section of a pork shoulder butt.
- **crayfish:** A freshwater crustacean that resembles a tiny lobster. Also known as a crawfish or crawdad.
- **crustacean:** A shellfish that has a hard, segmented shell protecting its soft flesh and that does not have an internal bone structure.
- **cuttlefish:** A translucent cephalopod with two tentacles, eight sucker-equipped arms, a hard internal cuttlebone, and large eyes at the base of its head.
- **double rib lamb chop:** A rib chop cut to a thickness equal to two standard rib chops.
- **drawn fish:** A fish that has had only the viscera removed.
- **dressed fish:** A fish that has been scaled and has had the viscera, gills, and fins removed.
- **drummette:** The innermost section of a wing located between the first wing joint and the shoulder.
- **drumstick:** The lower portion of a leg of poultry, located below the hip and above the knee joint.

- **Dungeness crab:** A Pacific crab with a sweet-tasting flesh.
- **English lamb chop:** A 2 inch thick fabricated cut taken along the entire length of the unsplit loin.
- **European oyster:** A variety of oyster with a relatively flat, cup-shaped shell and salty-sweet flavored, creamy-textured flesh.
- **fabricated cut:** A ready-to-cook cut that is packaged to certain size and weight specifications.
- **fatback:** The layer of fat that runs along the back of the hog.
- **fish:** Any of a classification of animal that has fins, gills, and an internal skeleton made of bones or cartilage.
- **fish fillet:** The lengthwise piece of flesh cut away from the backbone.
- **fish steak:** A cross section of a dressed fish.
- **flatfish:** Any thin, wide fish with both eyes located on one side of the head and a backbone that runs from head to tail through the lateral line of the body.
- **foie gras:** The fattened liver of a duck or goose.
- **frenching:** A method of removing the meat and fat from the end of a bone; generally applied to chops.
- **fryer/roaster turkey:** A male or female turkey that is less than three months old.
- **geoduck:** A very large Pacific clam with a meaty siphon that protrudes from its shell.
- **giblets:** The name for the grouping of the neck, heart, gizzard, and liver of a bird.
- **glazing:** The process of covering an item with water to form a protective coating of ice before the item is frozen.
- **grain-fed beef:** Meat from cattle that were grain-fed in confined feeding operations for 90 days to 1 year.
- **grass-fed beef:** Meat from cattle that were raised on grass or hay with little or no special feed.
- **ground poultry:** Ground fabricated cuts of poultry.
- **hard-shell clam:** An Atlantic clam with a blue-grey shell that contains a chewy flesh. Also known as a quahog.
- **headcheese:** The spiced, pressed, and jellied meat from the head of a hog.
- **honeycomb tripe:** The lining of the second stomach found in cattle.
- **individually quick-frozen (IQF):** A designation for products preserved using a method in which each item is glazed with a thin layer of water and frozen individually.
- **jowl:** Meat from the cheek of a hog.

KEY TERMS (continued)

- **king crab:** The largest-sized variety of crab, typically weighing between 6–20 lb and measuring as much as 10 feet from the tip of one leg to the tip of the opposite leg.
- **lamb:** The meat from slaughtered sheep that are less than a year old.
- **lamb back:** A lamb rack and loin that are still joined.
- **lamb bracelet:** A hotel rack with the breast still attached.
- **lamb breast:** A thin, flat, primal cut of lamb that contains the breastbone, the tips of the rib bones, and cartilage that is located under the shoulder and ribs.
- **lamb crown roast:** A hotel rack containing sixteen ribs with the bones frenched, notched, and tied to create a circle that resembles a crown.
- **lamb foresaddle:** The front half of the carcass consisting of the primal shoulder, rack, and breast cuts.
- **lamb hindsaddle:** The rear half of the carcass consisting of the loin and leg.
- **lamb hotel rack:** A lamb back that remains joined along the backbone.
- **lamb loin:** A primal cut located between the rack and leg that includes the 13th rib, the loin eye muscle, the center section of the tenderloin, the strip loin, and some flank meat.
- **lamb rack:** Eight rib bones located between the shoulder and loin of a lamb.
- **lamb riblet:** A rectangular strip of meat cut from the lamb breast that contains part of a rib bone.
- **lamb shank:** A cut of lamb that contains the upper foreshank bones.
- **lamb shoulder:** The first four rib bones of each side and the arm and neck bones.
- **langoustine:** A small lobster that resembles a very large shrimp, except it has very long front arms with long, thin claws. Also known as a Norway lobster or a Dublin Bay prawn.
- **leg of lamb:** A primal cut of lamb that contains the last portion of the backbone, hip bone, aitchbone, round bone, hindshank, and tailbone.
- **leg quarter:** A thigh, a drumstick, and a portion of the back.
- **lobster:** A saltwater crustacean with a brown to bluish-black external shell and two large claws.
- **Maine lobster:** A large clawed lobster with a dark bluish-green shell, two heavy claws, and eight slender legs.
- **mature duck:** A duck that is more than six months old.
- **mature turkey:** A turkey that is more than 15 months old.
- **mollusk:** A shellfish with a soft, nonsegmented body.
- **mussel:** A freshwater or saltwater bivalve with whisker-like threads that extend outside the shell to allow the animal to attach to items for protection.

- **noisette:** A small, round, boneless medallion of meat.
- **octopus:** A gray cephalopod with eight sucker-equipped arms, a birdlike beak, well-developed vision, and no internal or external shell.
- **offal:** An edible part of an animal that is not part of a primal cut.
- **oxtail:** The tail from a cattle carcass.
- **oyster:** A saltwater bivalve with a very rough shell that is coated with calcium deposits.
- **Pacific oyster:** A variety of large oyster that has fragile, curvy shells and a briny, sweet, and mild-tasting flesh. Also known as a Japanese oyster.
- **pancetta:** Unsmoked pork belly that has been cured in salt and spices, such as nutmeg and pepper, and then dried for a few months. Also known as Italian bacon.
- **pan-dressed fish:** A dressed fish that has had its head removed. Also known as a headed and gutted (H&G) fish.
- **picnic:** A cut of pork fabricated from the upper part of the foreleg that includes a portion of the shoulder.
- **picnic shoulder:** A primal cut of pork that is the lower half of the shoulder of a hog.
- **pork belly:** A fatty slab of meat and skin from the side and belly of a hog.
- **pork leg:** A primal cut of pork that is composed of the hind thigh and buttock of a hog. Also known as a ham.
- **pork loin:** A primal cut that extends along the backbone, from about the second rib, through the rib and loin area of a hog.
- **pork shoulder butt:** A square, compact area of the shoulder located just above the front legs of a hog. Also known as Boston butt.
- **pork spareribs:** The long, narrow ribs and breastbone of a hog.
- **pork tenderloin:** A fairly long, tapered strip of lean meat taken from the underside of the loin.
- **poultry:** The collective term for various kinds of birds that are raised for human consumption.
- **poultry half:** A full half-length of a bird split down the breast and spine.
- **poultry leg:** A drumstick and thigh.
- **poultry tenderloin:** The inner pectoral muscle that runs alongside the breastbone of a bird.
- **primal cut:** A large cut from a whole or a partial carcass.
- **prosciutto:** A type of dry-cured Italian ham.
- **quarter of beef:** A side of beef that has been divided into two parts between the 12th and 13th ribs.
- **rib eye:** A large, eye-shaped muscle within the rib that is a continuation of the sirloin muscle.

KEY TERMS (continued)

- **roaster:** A young male or female chicken that is 2–3 months old and has a ready-to-cook carcass weight of 5 lb or more.
- **roaster duckling:** A duck that is less than four months old.
- **roundfish:** Any fish with a cylindrical body, an eye located on each side of the head, and a backbone that runs from head to tail in the center of the body.
- **scallop:** A bivalve with a fan-shaped shell and a cream-colored adductor muscle with a sweet, delicate flavor.
- **scallopini:** A small, ¼ inch thick slice of veal (generally leg meat) that is 2–3 inches in diameter.
- **sea scallop:** A large scallop with a coarse texture that is harvested from deep saltwater.
- **shellfish:** The classification for aquatic invertebrates that may or may not have a hard, external shell.
- **shucking:** The process of opening a bivalve.
- **side of beef:** A half of a carcass split along the backbone.
- **siphon:** A tubular organ that is used to draw in or eject fluids.
- **sleeper:** A lobster that is dying.
- **snow crab:** A crab that is similar to a king crab but smaller and available in greater supply.
- **soft-shell clam:** An Atlantic clam with a thin, brittle shell that breaks easily. Also known as a long-neck or steamer clam.
- **soft-shell crab:** A blue crab that has been harvested within 6 hours of molting, or shedding its shell in order to grow a larger shell.
- **spiny lobster:** A lobster that has spines covering its body and five slender legs on each side. Also known as a rock lobster.
- **squid:** A translucent, head-footed cephalopod that has two tentacles, eight sucker-equipped arms, two lateral fins, and a flat, internal cuttlebone.
- **steamship round roast:** The beef round with the shank and rump removed.
- **steer:** A male calf that has been castrated prior to reaching sexual maturity.
- **stewer:** A female chicken that is more than 10 months old. Also known as a stewing hen.
- **stone crab:** An Atlantic crab with a brownish-red shell and large claws of unequal size.
- **suckling pig:** A pig 4–6 weeks old that weighs 20–35 lb dressed.
- **surf clam:** A large species of Atlantic clam that can grow to 8 inches in size.
- **surimi:** A fish product made from a mixture of fish and/or shellfish and other ingredients.
- **sweetbreads:** The thymus glands of a calf, located in the neck.

- **tender:** A small strip of a poultry breast.
- **thigh:** The upper section of a leg of poultry located below the hip and above the knee joint.
- **tripe:** The muscular inner lining of a stomach of an animal, such as cattle or sheep.
- **trussing:** The process of tying the legs and wings of a bird tightly to the body to keep a compact shape.
- **univalve:** A mollusk that has a single solid shell and a single foot. Also known as a gastropod.
- **veal:** The flesh of calves, which are young cattle.
- **veal breast:** A thin, flat cut of meat located under the shoulder and ribs; contains the breastbone, tips of the rib bones, and cartilage.
- **veal cutlet:** A thin slice of veal.
- **veal foresaddle:** The front half of a calf carcass consisting of the primal shoulder, rack, breast, and shank cuts.
- **veal foreshank:** The upper portion of the front leg of a calf.
- **veal hindsaddle:** The rear half of a calf carcass consisting of the loin and leg.
- **veal leg:** A primal cut from the hind leg that contains the leg, sirloin, last portion of the backbone, pelvis (hip bone and aitchbone), round bone, hindshank, and tailbone.
- **veal loin:** A primal cut of a calf located between the primal rack and leg; includes the 12th and 13th rib, the loin eye muscle, the center section of the tenderloin, the strip loin, and flank meat.
- **veal rack:** A primal cut located between the shoulder and loin of a calf and containing seven rib bones.
- **veal shoulder:** A primal cut of a calf that contains the first four rib bones, some of the backbone, and a small amount of the arm and blade bones.
- **wheel:** The round center cut of a large fish from which steaks are cut.
- **whole fish:** The market form of a fish that is taken from the water and sold as is.
- **wing:** The tip, paddle, and drummette.
- **wing paddle:** The second section of a wing located between the two wing joints. Also known as a wing flat.
- **wing tip:** The outermost section of a wing.
- **yearling turkey:** A mature turkey that is less than 15 months old.
- **young goose:** A goose that is usually less than six months of age and weighs approximately 4–10 lb.
- **young turkey:** A male or female turkey that is less than eight months old.

Refer to CD-ROM for **Quick Quiz®** questions

Sustainability Corner

"Food miles" refers to the distance food travels from where it is produced to where it is consumed. Food miles are one factor used when assessing the environmental impact of food, including its impact on global warming. This type of metric is sometimes used as a carbon emission label on packaging.

Recent findings indicate that it is not only how far the food has travelled, but the method of travel that is important to consider. Food that is transported by road produces 60% of the world's food transport carbon emissions. Air transport produces 20%, and rail and sea transport produce 10% each. The concept of food miles is part of the broader issue of sustainability that deals with a large range of environmental issues, including local food. In purchasing from organic farms that are not local, the positive environmental effects of organic farming may be compromised by increased transportation.

Studies of the total carbon footprint of food production in the U.S. have shown transportation to be of minor importance, however, compared to the carbon emissions resulting from pesticide and fertilizer production and the fuel required by farm and food-processing equipment. It is therefore important to identify organic products that can be purchased in your area of the country and how you would procure them. Finding local suppliers of organic foods can be achieved by visiting neighborhood farms. Also, websites such as Sustainable Table provide valuable links to locally produced, high-quality, organic foods.

Learning which foods are sustainable may change from location to location, with many variables within the choices. Costs may be higher, but so is quality. Local foods often taste fresher, look more vibrant, and typically have greater value to the consumer. Buying local supports the region's economy, creates jobs, and ensures the survival of the small farmer. Also, your customer needs to know if you support local organic farmers. Statistics show that people are now choosing restaurants that practice sustainability, not just in recycling, but in all areas of the restaurant, from buying only sustainable seafood to offering environmentally safe coffees and local wines and beers. The disadvantages are that local foods may only be available during certain times of the year and their supplies may be limited.

Sustainably raised animals are often treated humanely, permitted to carry out their natural behaviors, and usually produce healthier and better tasting proteins. Meats that come from local farms also take less travel time to reach your plate. There are other benefits as well. Studies show that grass-fed beef has two to six times more omega-3s than factory-farmed beef. Fewer antibiotics are used because of less overcrowding, and therefore there is less spread of disease. Waste is also controlled and used as fertilizer instead of being left untreated to pollute the water.

Not all foods purchased by your establishment can come from a local farmer. But when possible, your decisions can foster practices that sustain the health of the soil, the environment, communities, and the people producing and eating the food.

Organic Food Resources

1. ___ is the most common kind of poultry raised for consumption.
   A. Chicken
   B. Duck
   C. Turkey
   D. Goose

2. A ___ is a female chicken that is more than 10 months old.
   A. capon
   B. Cornish game hen
   C. broiler/fryer
   D. stewer

3. A ___ is a duck that is less than four months old.
   A. broiler/fryer duckling
   B. roaster duckling
   C. mature duck
   D. yearling duck

4. Foie gras is the fattened ___ of a duck or goose.
   A. kidney
   B. liver
   C. thymus
   D. tripe

5. The ___ is the name for the grouping of the neck, heart, gizzard, and liver of a bird.
   A. foresaddle
   B. hindsaddle
   C. giblets
   D. WOG

6. A ___ is any thin, wide fish with both eyes located on one side of the head and a backbone that runs from head to tail through the lateral line of the body.
   A. flatfish
   B. roundfish
   C. cephalopod
   D. mollusk

7. A ___ fish is a fish that has only the viscera removed.
   A. dressed
   B. pan-dressed
   C. H&G
   D. drawn

8. Fish steaks from large fish are fabricated from a ___.
   A. loin
   B. fillet
   C. wheel
   D. portion

9. Grade ___ foie gras is round and firm with no blemishes.
   A. 1A
   B. A
   C. B
   D. C

10. A blood line is a dark-red, almost black, muscle that runs along the ___ of a fish.
    A. back
    B. belly
    C. viscera
    D. lateral line

11. A ___ is a shellfish that has a hard segmented shell protecting its soft flesh and does not have an internal bone structure.
    A. crustacean
    B. mollusk
    C. univalve
    D. bivalve

12. A soft-shell crab is a ___ crab that is harvested within 6 hours of molting, or shedding its shell.
    A. snow
    B. Dungeness
    C. blue
    D. stone

13. A ___ is a very large Pacific clam with a meaty siphon that protrudes from its shell.
    A. gastropod
    B. geoduck
    C. surf clam
    D. quahog

14. A ___ is any of a variety of mollusks that do not have an external shell.
    A. univalve
    B. bivalve
    C. gastropod
    D. cephalopod

15. A ___ cut is a large cut from a whole or a partial carcass.
    A. primal
    B. fabricated
    C. side
    D. top

16. The ___ is the largest primal cut of beef and contains the first five rib bones, some of the backbone, and a small amount of the arm and blade bones.
    A. chuck
    B. sirloin
    C. round
    D. rib

17. A(n) ___ is an edible part of an animal that is not part of a primal cut.
    A. short plate
    B. flank
    C. offal
    D. rib

18. Sweetbreads are the ___ of a calf.
    A. brains
    B. thymus glands
    C. liver
    D. kidneys

19. The ___ is the largest primal cut of pork.
    A. picnic shoulder
    B. shoulder butt
    C. belly
    D. leg

20. A lamb bracelet is a hotel rack with the ___ still attached.
    A. back
    B. breast
    C. loin
    D. leg

21. Describe the procedure for trussing whole poultry.

_____

_____

_____

_____

_____

22. Describe the procedure for filleting flatfish.

_____

_____

_____

_____

_____

23. Describe the procedure for tying a boneless pork roast.

_____

_____

_____

_____

_____

_____

24. Describe the procedure for frenching racks of lamb.

_____

_____

_____

_____

_____

_____

SKETCHING EXERCISE

25. Sketch a beef carcass and identify the locations of all of the primal cuts.